Theory of Fuzzy Differential Equations and Inclusions

SERIES IN MATHEMATICAL ANALYSIS AND APPLICATIONS

Series in Mathematical Analysis and Applications (SIMAA) is edited by Ravi P. Agarwal, Florida Institute of Technology, USA and Donal O'Regan, National University of Ireland, Galway, Ireland.

The series is aimed at reporting on new developments in mathematical analysis and applications of a high standard and of current interest. Each volume in the series is devoted to a topic in analysis that has been applied, or is potentially applicable, to the solutions of scientific, engineering and social problems.

Volume 1
Method of Variation of Parameters for Dynamic Systems
V. Lakshmikantham and S.G. Deo

Volume 2
Integral and Integrodifferential Equations: Theory, Methods and Applications
edited by Ravi P. Agarwal and Donal O'Regan

Volume 3
Theorems of Leray-Schauder Type and Applications
Donal O'Regan and Radu Precup

Volume 4
Set Valued Mappings with Applications in Nonlinear Analysis
edited by Ravi P. Agarwal and Donal O'Regan

Volume 5
Oscillation Theory for Second Order Dynamic Equations
Ravi P. Agarwal, Said R. Grace and Donal O'Regan

Volume 6
Theory of Fuzzy Differential Equations and Inclusions
V. Lakshmikantham and R.N. Mohapatra

Volume 7
Monotone Flows and Rapid Convergence for Nonlinear Partial Differential Equations
V. Lakshmikantham and S. Köksal

This book is part of a series. The publisher will accept continuation orders which may be cancelled at any time and which provide for automatic billing and shipping of each title in the series upon publication. Please write for written details.

Theory of Fuzzy Differential Equations and Inclusions

V. Lakshmikantham

and

R. N. Mohapatra

CRC Press
Taylor & Francis Group
Boca Raton London New York

CRC Press is an imprint of the
Taylor & Francis Group, an **informa** business

A TAYLOR & FRANCIS BOOK

CRC Press
Taylor & Francis Group
6000 Broken Sound Parkway NW, Suite 300
Boca Raton, FL 33487-2742

First issued in paperback 2019

© 2003 V. Lakshmikantham and R. N. Mohapatra
CRC Press is an imprint of Taylor & Francis Group, an Informa business

No claim to original U.S. Government works

ISBN-13: 978-0-415-30073-5 (hbk)
ISBN-13: 978-0-367-39532-2 (pbk)

British Library Cataloguing in Publication Data
A catalogue record for this book is available from the British Library

Library of Congress Cataloging in Publication Data
A catalog record for this book has been requested

Visit the Taylor & Francis Web site at
http://www.taylorandfrancis.com

and the CRC Press Web site at
http://www.crcpress.com

Contents

Preface

In the mathematical modeling of real world phenomena, we encounter two inconveniences. The first is caused by the excessive complexity of the model. As the complexity of the system being modeled increases, our ability to make precise and yet relevant statements about its behavior diminishes until a threshold is reached beyond which precision and significance become almost mutually exclusive characteristics. As a result, we are either not able to formulate the mathematical model or the model is too complicated to be useful in practice.

The second inconvenience relates to the indeterminacy caused by our inability to differentiate events in a real situation exactly, and therefore to define instrumental notions in precise form. This indeterminacy is not an obstacle, when we use natural language, because its main property is the vagueness of its semantics and therefore capable of working with vague notions. Classical mathematics, on the other hand, cannot cope with such vague notions. It is therefore necessary to have some mathematical apparatus to describe vague and uncertain notions and thereby help to overcome the foregoing obstacles in the mathematical modeling of imprecise real world systems.

The rise and development of new fields such as general system theory, robotics, artificial intelligence and language theory, force us to be engaged in specifying imprecise notions. In 1965, Zadeh initiated the development of the modified set theory known as fuzzy set theory, which is a tool that makes possible the description of vague notions and manipulations with them. The basic idea of fuzzy set theory is simple and natural. A fuzzy set is a function from a set into a lattice or as a special case, into the interval $[0, 1]$. Using it, one can model the meaning of vague notions and also certain kinds of human reasoning. Fuzzy set theory and its applications have been extensively developed since the 1970s and industrial interest in fuzzy control has dramatically increased since 1990. There are several books dealing with these aspects.

When a real world problem is transformed into a deterministic initial value problem of ordinary differential equations, namely

$$x' = f(t, x), \quad x(t_0) = x_0, \qquad\qquad *$$

or a system of differential equations, we cannot usually be sure that the model is perfect. For example, the initial value may not be known exactly and the function f may contain uncertain parameters. If they are estimated through certain measurements, they are necessarily subject to errors. The analysis of the effect of these errors leads to the study of the qualitative behavior of the solutions of $(*)$. If the nature of errors is random, then we can discuss, instead of $(*)$, random differential equations with random initial data. However, if the underlying structure is not probabilistic, because of subjective choices, it would be natural to employ fuzzy differential equations. For the initiation of this aspect of fuzzy theory, the necessary calculus of fuzzy functions has also been investigated. Consequently, the study of the theory of fuzzy differential equations has recently been growing very rapidly and it is still in the initial stages. Nonetheless, there exists sufficient literature to warrant assembling the existing results in a unified way so as to understand and appreciate the intricacies involved in incorporating fuzziness into the theory of differential equations as well as to pave the way for further advancement of this important branch of differential equations as an independent discipline. It is with this spirit that we see the importance of the present monograph. Its aim is to present a systematic account of recent developments, describe the current state of the useful theory, show the essential unity achieved in the theory fuzzy differential equations or inclusions, and initiate several new extensions to other types of fuzzy dynamic systems.

In Chapter 1, we provide the preliminary material of fuzzy set theory providing necessary tools that are relevant for further development. Chapter 2 is dedicated to the description of the calculus of fuzzy functions. In Chapter 3, we devote our attention to investigate the basic theory of fuzzy differential equations. The extension of the Lyapunov-like theory of stability forms the content of Chapter 4. Chapter 5 investigates several new areas of investigation relative to fuzzy dynamic systems by providing some typical results so that further advancement is possible. Finally, in Chapter 6, we introduce fuzzy differential inclusions and investigate properties of solution sets, stability and periodicity in the new framework suggested by Hüllermeir. This new approach has the advantage of preserving the properties of solutions corresponding to differential equations without fuzziness. As we shall illustrate in Chapter 6, the original formulation based on the Hukuhara derivative totally changes the qualitative behavior of solutions when the initial condition is given more uncertainty by fuzzification. However, it can be preserved if the initial level sets are chosen suitably.

Some of the important features of the monograph are as follows:

(1) it is the first book that attempts to describe the theory of fuzzy differential equations;

(2) it incorporates the recent general theory (still in the pipeline) of fuzzy differential inclusions;

(3) it exhibits several new areas of study by providing initial apparatus for future development;

(4) it is a timely introduction to a subject that is growing rapidly because of its applicability to various new fields in engineering, computer science and social sciences.

Actually, the first five chapters of the monograph were written three years ago and because of various circumstances such as serious health problems and other unavoidable situations, the book could not be typed until now. This enormous delay turned out to be a blessing in disguise, since the new approach suggested by Hüllermeir, namely, developing the theory of fuzzy differential inclusions, is a better framework compared to the earlier one utilizing the Hukuhara derivative. We do hope that these two different approaches of considering fuzzy dynamic systems will generate other possible settings that may lead to a better understanding of incorporating fuzziness into various dynamic systems.

We are immensely grateful to Professors Hüllermier, Diamond, Nieto, and Seikkala for providing the material related to fuzzy differential inclusions and Mrs. Donn Miller-Kermani for typing the manuscript efficiently in a short time.

Chapter 1

Fuzzy Sets

1.1   Introduction

An exact description of any real world phenomenon is virtually impossible and one needs to accept this fact and adjust to it. The inexactness of the description is not a liability but is a blessing because it makes for greater efficiency. To specify imprecise or vague notions, Zadeh introduced the concept of fuzzy set theory. A fuzzy set is a membership function which describes the gradual transition from membership to nonmembership and is a subjective one. Spaces of such fuzzy sets are function spaces with special properties.

Since the monograph of Diamond and Kloeden [24] provides a good exposition of fuzzy set theory outlining the background and covering a broad aspect of topological properties of spaces, we have included minimal background material sufficient to deal with the theory of fuzzy differential equations and inclusions. In fact, the contents of Chapter 1 are adapted from their book.

Section 1.2 considers fuzzy sets, Zadeh's extension principle and the necessary spaces. Section 1.3 is devoted to the Hausdorff distance between subsets of $\mathcal{R}^n$ and its properties. Support functions form the content of Section 1.4. Theory of metric spaces of normal, upper semicontinuous, fuzzy convex fuzzy sets with compact support sets on the base space $\mathcal{R}^n$, is discussed in Section 1.5. This section includes the representation theorem which is also important in the development of fuzzy differential inclusions. Section 1.6 deals with the properties of metric space (E^n, d) proving its completeness and properties of the metric d. Finally, Section 1.7 provides notes and comments.

1.2   Fuzzy Sets

The idea of a fuzzy set was first proposed by Lotfi Zadeh in the 1960s, as a means of handling uncertainty that is due to imprecision or vagueness rather than to randomness.

Fuzzy sets are considered with respect to a nonempty base set X of elements of interest. The essential idea is that each element $x \in X$ is assigned a membership grade $u(x)$ taking values in $[0, 1]$, with $u(x) = 0$ corresponding to nonmembership, $0 < u(x) < 1$ to partial membership, and $u(x) = 1$ to full membership. According to Zadeh a *fuzzy subset* of X is a nonempty subset $\{(x, u(x)) : x \in X\}$ of $X \times [0, 1]$ for some function $u : X \to [0, 1]$. The function u itself is often used for the fuzzy set.

For instance, the function $u : \mathcal{R}^1 \to [0, 1]$ with

$$u(x) = \begin{cases} 0 & \text{if} \quad x \leq 1 \\ \frac{1}{99}(x - 1) & \text{if} \quad 1 < x \leq 100 \\ 1 & \text{if} \quad 100 < x \end{cases} \qquad (1.2.1)$$

provides an example of a fuzzy set of real numbers. There are of course many other reasonable choices of membership grade function. The only membership possibilities for an ordinary or *crisp* subset A of X are nonmembership and full membership. Such a set can thus be identified with a *fuzzy set* on X given by its characteristic function $\chi_A : X \to [0, 1]$, that is, with

$$\chi_A(x) = \begin{cases} 0 : & x \notin A \\ 1 : & x \in A. \end{cases} \qquad (1.2.2)$$

Metric spaces of fuzzy sets provide a convenient mathematical framework for diverse applications of fuzzy sets. They are essentially spaces of special kinds of functions from a base space X to $[0, 1]$, where X is a metric space.

The α-*level set* $[u]^\alpha$ of a fuzzy set u on X is defined as

$$[u]^\alpha = \{x \in X : u(x) \geq \alpha\} \quad \text{for each} \quad \alpha \in (0, 1], \qquad (1.2.3)$$

while its *support* $[u]^0$ is the closure in the topology of X of the union of all of the level sets, that is

$$[u]^0 = \overline{\bigcup_{\alpha \in (0,1]} [u]^\alpha}. \qquad (1.2.4)$$

The union, intersection and complement of fuzzy sets can be defined pointwise in terms of their membership grades. Consider a function $u : X \to [0, 1]$ as a fuzzy subset of a nonempty base space X and denote the totality of all such functions or fuzzy sets by $\mathcal{F}(X)$.

The *complement* u^c of $u \in \mathcal{F}(X)$, the *union* $u \vee v$ and the *intersection* $u \wedge v$ of $u, v \in \mathcal{F}(X)$ are defined, respectively, by

$$u^c(x) = 1 - u(x), \qquad\qquad (1.2.5)$$

$$u \vee v(x) = u(x) \vee v(x) := \max\{u(x), v(x)\}, \qquad\qquad (1.2.6)$$

$$u \wedge v(x) = u(x) \wedge v(x) := \min\{u(x), v(x)\} \qquad\qquad (1.2.7)$$

for each $x \in X$. Clearly u^c, $u \vee v$, $u \wedge v \in \mathcal{F}(X)$.

The *Zadeh extension principle* allows a crisp mapping $f : X_1 \times X_2 \to Y$, where X_1, X_2, and Y are nonempty sets, to be extended to a mapping on fuzzy sets

$$\tilde{f} : \mathcal{F}(X_1) \times \mathcal{F}(X_2) \to \mathcal{F}(Y)$$

where

$$\tilde{f}(u_1, u_2)(y) = \left\{ \begin{array}{r} \sup_{(x_1, x_2) \in f^{-1}(y)} u_1(x_1) \wedge u_2(x_2), \quad \text{if} \quad f^{-1}(y) \neq \emptyset, \\ 0, \quad \text{if} \quad f^{-1}(y) = \emptyset, \end{array} \right.$$
$$(1.2.8)$$

for $y \in Y$. Here $f^{-1}(y) = \{(x_1, x_2) \in X_1 \times X_2 : f(x_1, x_2) = y\}$ may be empty or contain one or more points. The obvious generalization holds for mappings defined on an N-tuple $X_1 \times \ldots \times X_N$ where $N \geq 1$, with the *wedge* operator being superfluous when $N = 1$.

The definitions of addition and scalar multiplication of fuzzy sets in $\mathcal{F}(X)$ involve the extension principle and require the base set X to be a linear space. For the *addition* of two fuzzy sets $u, v \in \mathcal{F}(X)$, the Zadeh extension principle is applied to the function $f : X \times X \to X$ defined by $f(x_1, x_2) = x_1 + x_2$ to give

$$(u \,\tilde{+}\, v)(x) = \sup_{x_1 + x_2 = x} u(x_1) \wedge v(x_2), \qquad\qquad (1.2.9)$$

for all $x \in X$, while for *scalar multiplication* of $u \in \mathcal{F}(X)$ by a nonzero scalar c, the function $f : X \to X$ defined by $f(x) = cx$ is extended to

$$\tilde{c}u(x) = u(x/c) \qquad\qquad (1.2.10)$$

for all $x \in X$. Obviously, both $u \,\tilde{+}\, v$ and $\tilde{c}u$ belong to $\mathcal{F}(X)$.

The totality of fuzzy sets $\mathcal{F}(X)$ on a base space X is often too broad and general to allow strong or specific enough results to be established, and therefore various restrictions are often imposed on the fuzzy sets. In particular, a fuzzy set $u \in \mathcal{F}(X)$ is called a *normal fuzzy set* if there exists at least one point $x_0 \in X$ for which $u(x_0) = 1$, so the 1-level set $[u]^1$ and hence every other level set $[u]^\alpha$ for $0 < \alpha < 1$ and the support $[u]^0$ of u are all nonempty subsets of X. For technical reasons, the level sets are often assumed to be compact and, when X is a linear space, also convex. In fact, the convexity of the level sets of a fuzzy set u is equivalent to its being a *fuzzy convex* fuzzy set, that is, satisfying

$$u(\lambda x_1 + (1 - \lambda)x_2) \geq u(x_1) \wedge u(x_2) \quad \text{for all} \quad x_1, x_2 \in X, \quad \lambda \in [0, 1].$$
$$(1.2.11)$$

In the case of *fuzzy numbers*, that is fuzzy sets $u : \mathcal{R} \to [0, 1]$, fuzzy convexity means that the level sets are *intervals*.

We shall consider the following three spaces of *nonempty* subsets of $\mathcal{R}^n$.

(i) $\mathcal{C}^n$ consisting of all nonempty *closed* subsets of $\mathcal{R}^n$,

(ii) $\mathcal{K}^n$ consisting of all nonempty *compact* subsets of $\mathcal{R}^n$,

(iii) $\mathcal{K}_C^n$ consisting of all nonempty *compact convex* subsets of $\mathcal{R}^n$.

We then have the strict inclusions

$$\mathcal{K}_C^n \subset \mathcal{K}^n \subset \mathcal{C}^n.$$

Recall that a nonempty subset A of $\mathcal{R}^n$ is convex if for all $a_1, a_2 \in A$ and all $\lambda \in [0, 1]$, the point

$$a = \lambda a_1 + (1 - \lambda)a_2 \qquad (1.2.12)$$

belongs to A. For any nonempty subset A of $\mathcal{R}^n$, we denote by $\text{co}A$ its *convex hull*, that is the totality of points a of the form (1.2.12) or, equivalently, the smallest convex subset containing A. Clearly

$$A \subseteq \text{co } A = \text{co}\,(\text{co}A) \qquad (1.2.13)$$

with $A = \text{co}A$ if A is convex. Moreover $\text{co}A$ is closed (compact) if A is closed (compact).

Let A and B be two nonempty subsets of $\mathcal{R}^n$ and let $\lambda \in \mathcal{R}$. We define the following Minkowski addition and scalar multiplication by

$$A + B = \{a + b : a \in A, b \in B\} \qquad (1.2.14)$$

and

$$\lambda A = \{\lambda a : a \in A\}. \qquad (1.2.15)$$

Then we have

Proposition 1.2.1. *The spaces* $\mathcal{C}^n$, $\mathcal{K}^n$ *and* $\mathcal{K}_C^n$ *are closed under the operations of addition and scalar multiplication.*

In fact, these two operations induce a *linear structure* on $\mathcal{C}^n$, $\mathcal{K}^n$ and $\mathcal{K}_C^n$ with zero element $\{0\}$. The structure is that of a cone rather than a vector space because, in general,

$$A + (-1)A \neq \{0\}.$$

Example 1.2.1. *Let* $A = [0, 1]$ *so that* $(-1)A = [-1, 0]$, *and therefore*

$$A + (-1)A = [0, 1] + [-1, 0] = [-1, 1].$$

Thus, adding -1 *times a set does not constitute a natural operation of subtraction. Instead, we define the Hukuhara difference* $A - B$ *of nonempty sets* A *and* B, *provided there exists a nonempty set* C *satisfying*

$$A = B + C. \tag{1.2.16}$$

Example 1.2.2. *From the preceding example,*

$$[-1, 1] - [-1, 0] = [0, 1] \quad and \quad [-1, 1] - [0, 1] = [-1, 0].$$

Clearly, $A - A = \{0\}$ *for all nonempty sets* A. *From* (1.2.16), *a necessary condition for the Hukuhara difference* $A - B$ *to exist is that some translate of* B *is a subset of* A,

$$B + \{c\} \subseteq A$$

for some $c \in \mathcal{R}^n$. *When it exists,* $A - B$ *is unique. However, that the Hukuhara difference need not exist is seen from the following example.*

Example 1.2.3. $\{0\} - [0, 1]$ *does not exist, since no translate of* $[0, 1]$ *can ever belong to the singleton set* $\{0\}$.

1.3 The Hausdorff Metric

Let x be a point in $\mathcal{R}^n$ and A a nonempty subset of $\mathcal{R}^n$. The *distance* $d(x, A)$ from x to A is defined by

$$d(x, A) = \inf\{\|x - a\| : a \in A\}. \tag{1.3.1}$$

Thus $d(x, A) = d(x, \bar{A}) \geq 0$ and $d(x, A) = 0$ if and only if $x \in \bar{A}$, the closure of A in $\mathcal{R}^n$.

We shall call the subset

$$S_\epsilon(A) = \{x \in \mathcal{R}^n : d(x, A) < \epsilon\} \tag{1.3.2}$$

an *ϵ-neighborhood of A*. Its closure is the subset

$$\bar{S}_\epsilon(A) = \{x \in \mathcal{R}^n : d(x, A) \le \epsilon\}. \tag{1.3.3}$$

In particular, we shall denote by $\bar{S}_1^n$ the *closed unit ball* in $\mathcal{R}^n$, that is

$$\bar{S}_1^n = \bar{S}_1(\{0\}) \tag{1.3.4}$$

which is obviously a compact subset of $\mathcal{R}^n$. Note that

$$\bar{S}_\epsilon(A) = A + \epsilon\bar{S}_1^n \tag{1.3.5}$$

for any $\epsilon > 0$ and any nonempty subset A of $\mathcal{R}^n$. We shall for convenience sometimes write $S(A, \epsilon)$ and $\bar{S}(A, \epsilon)$ for $S_\epsilon(A)$ and $\bar{S}_\epsilon(A)$.

Now let A and B be nonempty subsets of $\mathcal{R}^n$. We define the *Hausdorff separation* of B from A by

$$d_H^*(B, A) = \sup\{d(b, A) : b \in B\} \tag{1.3.6}$$

or, equivalently,

$$d_H^*(B, A) = \inf\{\epsilon > 0 : B \subseteq A + \epsilon\bar{S}_1^n\}. \tag{1.3.7}$$

We have $d_H^*(B, A) \ge 0$ with $d_H^*(B, A) = 0$ if and only if $B \subseteq \bar{A}$. Also, the *triangle inequality*

$$d_H^*(B, A) \le d_H^*(B, C) + d_H^*(C, A)$$

holds for all nonempty subsets A, B and C of $\mathcal{R}^n$. In general, however

$$d_H^*(A, B) \ne d_H^*(B, A).$$

We define the *Hausdorff distance* between nonempty subsets A and B of $\mathcal{R}^n$ by

$$d_H(A, B) = \max\{d_H^*(A, B), d_H^*(B, A)\}, \tag{1.3.8}$$

which is symmetric in A and B. Consequently,

(a) $d_H(A, B) \ge 0$ with $d_H(A, B) = 0$ if and only if $\bar{A} = \bar{B}$;
(b) $d_H(A, B) = d_H(B, A)$;
(c) $d_H(A, B) \le d_H(A, C) + d_H(C, B),$

$$\tag{1.3.9}$$

for any nonempty subsets A, B and C of $\mathcal{R}^n$. If we restrict our attention to nonempty closed subsets of $\mathcal{R}^n$, we find that the Hausdorff distance (1.3.8) is a metric, known as the *Hausdorff metric* [37]. Thus $(\mathcal{C}^n, d_H)$ is a metric space. In fact, we have

Proposition 1.3.1. *$(\mathcal{C}^n, d_H)$ is a complete separable metric space in which $\mathcal{K}^n$ and $\mathcal{K}_C^n$ are closed subsets. Hence, $(\mathcal{K}^n, d_H)$ and $(\mathcal{K}_C^n, d_H)$ are also complete separable metric spaces.*

The following properties of the Hausdorff metric will be useful in the sequel.

Proposition 1.3.2. *If A, A', B, $B' \in \mathcal{K}^n$, then*

$$d_H(tA, tB) = t d_H(A, B) \quad \text{for all} \quad t \geq 0, \tag{1.3.10}$$

$$d_H(A + B, A' + B') \leq d_H(A, A') + d_H(B, B'), \tag{1.3.11}$$

$$d_H(\text{co } A, \text{co } B) \leq d_H(A, B). \tag{1.3.12}$$

Proposition 1.3.3. *If A, $B \in \mathcal{K}_C^n$ and $C \in \mathcal{K}^n$ then*

$$d_H(A + C, B + C) = d_H(A, B). \tag{1.3.13}$$

We define the *magnitude* of a nonempty subset of A of $\mathcal{R}^n$ by

$$\|A\| = \sup\{\|a\| : a \in A\} \tag{1.3.14}$$

or, equivalently,

$$\|A\| = d_H(\{0\}, A). \tag{1.3.15}$$

Here, $\|A\|$ is finite and the supremum in (1.3.14) is attained when $A \in \mathcal{K}^n$. From (1.3.10) it obviously follows that

$$\|tA\| = t\|A\| \quad \text{for all} \quad t \geq 0. \tag{1.3.16}$$

Moreover, (1.3.9) and (1.3.15) yield

$$|\|A\| - \|B\|| \leq d_H(A, B) \tag{1.3.17}$$

for all $A, B \in \mathcal{K}^n$.

We say that a subset $\mathcal{U}$ of $\mathcal{K}^n$ (or $\mathcal{K}_C^n$) is *uniformly bounded* if there exists a finite constant $c(\mathcal{U})$ such that

$$\|A\| \leq c(\mathcal{U}) \quad \text{for all} \quad A \in \mathcal{U}. \tag{1.3.18}$$

We then have the following simple characterization of compactness.

Proposition 1.3.4. *A nonempty subset $\mathcal{A}$ of the metric space $(\mathcal{K}^n, d_H)$, or $(\mathcal{K}_C^n, d_H)$, is compact if and only if it is closed and uniformly bounded.*

Set inclusion induces a partial ordering on $\mathcal{K}^n$. Write $A \leq B$ if and only if $A \subseteq B$, where $A, B \in \mathcal{K}^n$. Then

$$\mathcal{L}(B) = \{A \in \mathcal{K}^n : B \leq A\}, \quad \mathcal{U}(B) = \{A \in \mathcal{K}^n : A \leq B\} \qquad (1.3.19)$$

are closed subsets of $\mathcal{K}^n$ for any $B \in \mathcal{K}^n$. In fact, from Proposition 1.3.4, $\mathcal{U}(B)$ is a compact subset of $\mathcal{K}^n$.

Proposition 1.3.5. *$\mathcal{U}(B)$ is a compact subset of $\mathcal{K}^n$.*

This assertion remains true with $\mathcal{K}^n_C$ replacing $\mathcal{K}^n$ everywhere.

Sequences of nested subsets in $(\mathcal{K}^n, d_H)$ have the following useful intersection and convergence properties.

Proposition 1.3.6. *Let $\{A_j\} \subset \mathcal{K}^n$ satisfy*

$$\ldots \subseteq A_j \subseteq \ldots \subseteq A_2 \subseteq A_1.$$

Then $A = \cap_{j=1}^{\infty} A_j \in \mathcal{K}^n$ and

$$d_H(A_n, A) \to 0 \quad as \quad n \to \infty. \qquad (1.3.20)$$

On the other hand, if $A_1 \subseteq A_2 \subseteq \ldots \subseteq A_j \subseteq \ldots$ and $A = \cup_{j=1}^{\infty} A_j \in \mathcal{K}^n$, then (1.3.20).

1.4   Support Functions

Let A be a nonempty subset of $\mathcal{R}^n$. The support function of A is defined for all $p \in \mathcal{R}^n$ by

$$s(p, A) = \sup\{\langle p, a \rangle : a \in A\}, \qquad (1.4.1)$$

which may take the value $+\infty$ when A is unbounded. However, when A is a compact, convex subset of $\mathcal{R}^n$ the supremum is always attained and the support function $s(\cdot, A) : \mathcal{R}^n \to \mathcal{R}$ is well defined. Indeed,

$$|s(p, A)| \leq \|A\| \|p\| \qquad (1.4.2)$$

for all $p \in \mathcal{R}^n$ and

$$|s(p, A) - s(q, A)| \leq \|A\| \|p - q\| \qquad (1.4.3)$$

for all $p, q \in \mathcal{R}^n$. In addition, for all $p \in \mathcal{R}^n$

$$s(p, A) \leq s(p, B) \quad \text{if} \quad A \subseteq B \qquad (1.4.4)$$

and

$$s(p, \text{co}(A \cup A)) \leq \max\{s(p, A), s(p, B)\}. \qquad (1.4.5)$$

The support function $s(p, A)$ is uniquely paired to the subset A in $\mathcal{K}_C^n$ in the sense that $s(p, A) = s(p, B)$ for all $p \in \mathcal{R}^n$ if and only if $A = B$ when A and B are restricted to $\mathcal{K}_C^n$. It also preserves set addition and nonnegative scalar multiplication. That is, for all $p \in \mathcal{R}^n$,

$$s(p, A + B) = s(p, A) + s(p, B) \qquad (1.4.6)$$

which, in particular, reduces to

$$s(p, A + \{x\}) = s(p, A) + \langle p, x \rangle \qquad (1.4.7)$$

for any $x \in \mathcal{R}^n$, and

$$s(p, tA) = ts(p, A), \quad t \geq 0. \qquad (1.4.8)$$

The Hausdorff metric is related to the support function for subsets A, $B \in \mathcal{K}_C^n$ since we have

$$d_H(A, B) = \sup\{|s(p, A) - s(p, B)| : p \in S^{n-1}\} \qquad (1.4.9)$$

where $S^{n-1} = \{x \in \mathcal{R}^n : \|x\| = 1\}$ is the unit sphere in $\mathcal{R}^n$.

Let $C(S^{n-1})$ denote the Banach space of continuous functions $f : S^{n-1} \to \mathcal{R}$ with the supremum norm

$$\|f\| = \sup\{\|f(p)\| : p \in S^{n-1}\}.$$

One can use the support function to embed the metric space $(\mathcal{K}_C^n, d_H)$ isometrically as a positive cone in $C(S^{n-1})$. For this, define $j : \mathcal{K}_C^n \to C(S^{n-1})$ by $j(A)(\cdot) = s(\cdot, A)$ for each $A \in \mathcal{K}_C^n$. From the properties of the support function, j is a univalent mapping satisfying

$$j(A + B) = j(A) + j(B) \qquad (1.4.10)$$

and

$$j(tA) = tj(A), \quad t \geq 0, \qquad (1.4.11)$$

with

$$\|j(A) - j(B)\| = d_H(A, B) \qquad (1.4.12)$$

for all $A, B \in \mathcal{K}_C^n$. The desired positive cone is the image $j(\mathcal{K}_C^n)$ in $C(S^{n-1})$. Obviously j is continuous, as is its inverse $j^{-1} : j(\mathcal{K}_C^n) \to \mathcal{K}_C^n$.

For a fixed $A \in \mathcal{K}_C^n$, $s(p, A)$ is positively homogeneous

$$s(tp, A) = ts(p, A), \quad t \geq 0 \tag{1.4.13}$$

for all $p \in \mathcal{R}^n$, and *subadditive*:

$$s(p_1 + p_2, A) \leq s(p_1, A) + s(p_2, A) \tag{1.4.14}$$

for all $p_1, p_2 \in \mathcal{R}^n$. Moreover, combining (1.4.13) and (1.4.14) we see that $s(\cdot, A)$ is a convex function, that is, it satisfies

$$s(\lambda p_1 + (1 - \lambda)p_2, A) \leq \lambda s(p_1, A) + (1 - \lambda)s(p_2, A) \tag{1.4.15}$$

for all $p_1, p_2 \in \mathcal{R}^n$ and $\lambda \in [0, 1]$.

The nonempty compact convex subsets of $\mathcal{R}^n$ are uniquely characterized by such functions.

Proposition 1.4.1. *For every continuous, positively homogeneous and subadditive function $s : \mathcal{R}^n \to \mathcal{R}$ there exists a unique nonempty compact convex subset*

$$A = \{x \in \mathcal{R}^n : \langle p, x \rangle \leq s(p) \quad for \ all \quad p \in \mathcal{R}^n\}$$

which has s as its support function.

1.5 The Space E^n

Recall that a fuzzy subset of $\mathcal{R}^n$ is defined in terms of a membership function which assigns to each point $x \in \mathcal{R}^n$ a grade of membership in the fuzzy set. Such a membership function

$$u : \mathcal{R}^n \to I = [0, 1]$$

is used to denote the corresponding fuzzy set. Denote by $\mathcal{F}^n$ the set of all fuzzy sets of $\mathcal{R}^n$.

For each $\alpha \in (0, 1]$ the α-*level set* $[u]^\alpha$ of a fuzzy set u is the subset of points $x \in \mathcal{R}^n$ with membership grade $u(x)$ of at least α, that is

$$[u]^\alpha = \{x \in \mathcal{R}^n : u(x) \geq \alpha\}.$$

The *support* $[u]^0$ of a fuzzy set is then defined as the closure of the union of all its level sets, that is

$$[u]^0 = \overline{\bigcup_{\alpha \in (0,1]} [u]^\alpha}.$$

Let $P_k(\mathcal{R}^n)$ denote the family of all nonempty compact convex subsets of $\mathcal{R}^n$. Define addition and scalar multiplication in $P_k(\mathcal{R}^n)$ by

$$A + B = [z : z = x + y, x \in A \quad \text{and} \quad y \in B]$$

for all $A, B \in P_k(\mathcal{R}^n)$ and

$$\lambda A = [z : z = \lambda x, x \in A]$$

for all $A \in P_k(\mathcal{R}^n)$. From Rådstrom [100], we know that $P_k(\mathcal{R}^n)$ forms a commutative semigroup under addition which satisfies the cancelation law. Further, if $\alpha, \beta \in \mathcal{R}$ and $A, B \in P_k(\mathcal{R}^n)$, then

$$\alpha(A + B) = \alpha A + \alpha B, \quad \alpha(\beta A) = (\alpha \beta) A, \quad 1A = A,$$

and for $\alpha, \beta > 0$, $(\alpha + \beta)A = \alpha A + \beta A$.

Let us denote by E^n the space of all fuzzy subsets u of $\mathcal{R}^n$ which satisfy the assumptions

(1) u maps $\mathcal{R}^n$ onto $I = [0, 1]$;

(2) $[u]^0$ is a bounded subset of $\mathcal{R}^n$;

(3) $[u]^\alpha$ is a compact subset of $\mathcal{R}^n$ for all $\alpha \in I$;

(4) u is fuzzy convex, that is

$$u(\lambda x + (1 - \lambda)y) \geq \min[u(x), u(y)]$$

for any $\lambda \in [0, 1]$.

Let u be fuzzy convex and $x, y \in [u]^\alpha$ for some $\alpha \in (0, 1]$. Hence $u(x) \geq \alpha$ and $u(y) \geq \alpha$. Then

$$u(\lambda x + (1 - \lambda)y) \geq \min[u(x), u(y)] \geq \alpha$$

for any $\lambda \in [0, 1]$ and therefore $\lambda x + (1 - \lambda)y \in [u]^\alpha$. As a result, $[u]^\alpha$ is a convex subset of $\mathcal{R}^n$ for any $\alpha \in (0, 1]$. The support $[u]^0$ is also convex, which follows from the fact that

$$d_H([u]^\alpha, [u]^0) \to \quad \text{as} \quad \alpha \to 0^+$$

and the completeness of the metric space $(\mathcal{K}_C^n, d_H)$. Thus we have:

Lemma 1.5.1. *If u is fuzzy convex, then $[u]^\alpha$ is convex for each $\alpha \in I$.*

We can now prove the following representation theorem.

Theorem 1.5.1. *If $u \in E^n$, then the following hold:*

$$[u]^\alpha \in P_k(\mathcal{R}^n) \quad for\ all \quad 0 \le \alpha \le 1, \tag{1.5.1}$$

$$[u]^{\alpha_2} \subset [u]^{\alpha_1} \quad for \quad 0 \le \alpha_1 \le \alpha_2 \le 1. \tag{1.5.2}$$

If (α_k) is a nondecreasing sequence converging to $\alpha > 0$, then

$$[u]^\alpha = \bigcap_{k \ge 1} [u]^{\alpha_k}. \tag{1.5.3}$$

Conversely, if $\{A^\alpha : 0 \le \alpha \le 1\}$ is a family of subsets of $\mathcal{R}^n$ satisfying (1.5.1)–(1.5.3) then there exists a $u \in E^n$ such that

$$[u]^\alpha = A^\alpha \quad for \quad 0 < \alpha \le 1$$

and

$$[u]^0 = \mathrm{cl} \bigcup_{0 < \alpha \le 1} A^\alpha \subset A^0.$$

Proof.   If $u \in E^n$, we have from the definition

$$[u]^\beta \subseteq [u]^\alpha \subseteq [u]^0,$$

where $[u]^0 = \mathrm{cl} \cup_{\alpha \in (0,1]} [u]^\alpha$. Since u is normal, u maps $\mathcal{R}^n$ onto $I = [0,1]$. Also $[u]^\alpha$ is a compact subset of $\mathcal{R}^n$ for all $\alpha \in I$ and for any nondecreasing sequence $\alpha_i \to \alpha$ in I, $[u]^\alpha = \cap_{i \ge 1} [u]^{\alpha_i}$. We also know u and $[u]^\alpha$ are convex by Lemma 1.5.1. This proves the first part of Theorem 1.5.1. We next prove the converse. For $x \in A^0$, define $I_x = [\alpha \in I : x \in A^\alpha]$ and let $\alpha_0 = \sup I_x$. We claim $I_x = [0, \alpha_0]$. If $\alpha_0 = 0$, there is nothing to prove and hence suppose that $\alpha_0 > 0$ and let $\beta \in (0, \alpha_0)$. Then there exists $\beta_1 \in [\beta, \alpha_0)$ such that $\beta_1 \in I_x$. Thus $x \in A^{\beta_1}$ which implies by (1.5.2), $x \in A^\beta$ and $\beta \in I_x$. By definition, $0 \in I_x$ and hence $[0, \alpha_0) \subseteq I_x$.

Now let α_i converge to α_0 in I_x monotonically. Then $x \in A^{\alpha_i}$ for each $i = 1, 2, \ldots$ and therefore by (1.5.3), $x \in A^{\alpha_0}$. Thus $\alpha_0 \in I_x$ and $[0, \alpha_0] \subseteq I_x$. Finally, since $\beta \in I_x$ implies that $\beta \le \alpha_0$, we have $I_x \subseteq [0, \alpha_0]$. Hence $I_x = [0, \alpha_0]$ as was asserted.

Let $\alpha \in [0, 1]$. If $x \in [u]^\alpha$, then $u(x) \ge \alpha > 0$ and so $x \in A^0$ and $u(x) = \sup I_x = \alpha_0 \ge \alpha$. Hence $x \in A^{\alpha_0}$ and consequently by (1.5.2) $x \in A^\alpha$, that is, $[u]^\alpha \subseteq A^\alpha$. Conversely, if $x \in A^\alpha$, then $u(x) = \sup I_x = \alpha_0 \ge \alpha$ and hence $x \in [u]^\alpha$. This implies $A^\alpha \subseteq [u]^\alpha$. Combining the results, we obtain $[u] = A^\alpha$ for all $\alpha \in (0,1]$.

Defining $[u]^0$ as above, we find that u maps $\mathcal{R}^n$ onto I and is upper semicontinuous (usc) since $[u]^\alpha$ are closed. Furthermore $[u]^0$ is compact and hence bounded. It is also convex. Thus u satisfies the requirements except convexity. To prove convexity, let $x, y \in [u]^\alpha$ with $\min[u(x), u(y)] = \gamma \geq \alpha$. Then, $x, y \in [u]^\gamma$, which is convex and so $\lambda x + (1 - \lambda)y \in [u]^\gamma$ for any $\lambda \in [0,1]$. Hence

$$u(\lambda x + (1 - \lambda)y) \geq \gamma = \min[u(x), u(y)],$$

and therefore u is fuzzy convex. Hence $u \in E^n$ and the proof is complete.

An advantage of only requiring $u \in E^n$ to be upper semicontinuous and not continuous is that the nonempty compact subsets of $\mathcal{R}^n$ can then also be included in E^n by means of their characteristic functions. By a straightforward application of Theorem 1.5.1, we obtain

Proposition 1.5.1. *If $A \in \mathcal{K}_C^n$, then $\chi_A \in E^n$.*

We shall also need the following result. See Castaing and Valadier [9] for a proof.

Theorem 1.5.2. *Let $\{A_k\}$ be a sequence in $P_k(\mathcal{R}^n)$ converging to A. Further, let $d_H(A_k, A) \to 0$ as $k \to \infty$. Then*

$$A = \bigcap_{k \geq 1} \{\mathrm{cl} U_{m \geq k} A_m\}.$$

In the context of fuzzy sets we call a subset of $\mathcal{R}^n$, or more precisely its characteristic function χ_A, a *crisp subset* of $\mathcal{R}^n$.

The *endograph* end(u) of a fuzzy set $u \in E^n$ is defined as

$$\mathrm{end}(u) = \{(x, \alpha) \in \mathcal{R}^n \times I : u(x) \leq \alpha\}. \tag{1.5.4}$$

It is a nonempty closed subset of $\mathcal{R}^n \times I$. Restricting to those points that lie above the support set, we obtain the *supported endograph*, or *sendograph* for short, of u

$$\mathrm{send}(u) = \mathrm{end}(u) \cap ([u]^0 \times I), \tag{1.5.5}$$

which is a nonempty compact subset of $\mathcal{R}^n \times I$. In fact,

$$\mathrm{send}(u) = \bigcup \{[u]^\alpha \times \{\alpha\} : \alpha \in I\}.$$

Example 1.5.1. *Let $A \in \mathcal{K}_C^n$, so $\chi_A \in E^n$. Then $[\chi_A] = \alpha$ for all $\alpha \in I$,*

$$end(\chi_A) = (\mathcal{R}^n \times \{0\}) \bigcup (A \times I)$$

and

$$send(\chi_A) = A \times I.$$

We shall define addition and scalar multiplication of fuzzy sets in E^n levelsetwise, that is, for $u, v \in E^n$ and $c \in \mathcal{R} \backslash \{0\}$

$$[u + v]^\alpha = [u]^\alpha + [v]^\alpha \tag{1.5.6}$$

and

$$[cu]^\alpha = c[u]^\alpha \tag{1.5.7}$$

for each $\alpha \in I$.

Proposition 1.5.2. *E^n is closed under addition (1.5.6) and scalar multiplication (1.5.7).*

Proof.   We apply Theorem 1.5.1 to the families of subsets $\{[u + v]^\alpha : \alpha \in I\}$ and $\{[cu]^\alpha\}$. Properties (1.5.1) and (1.5.2) follow from those for $\{[u]^\alpha : \alpha \in I\}$ and $\{[v]^\alpha : \alpha \in I\}$, definitions (1.5.6) and (1.5.7), and the closedness of $\mathcal{K}_C^n$ under set addition and scalar multiplication. Now let $\{\alpha_i\}$ be a nondecreasing sequence in I with $\alpha_i \uparrow \alpha$ in I. Then, by (1.5.6) and (1.5.7), by property (1.5.3) for $\{[u]^\alpha : \alpha \in I\}$ and $\{[v]^\alpha : \alpha \in I\}$, and by Proposition 1.3.2,

$$\begin{aligned} d_H([u + v]^{\alpha_i}, [u + v]^\alpha) &= d_H([u]^{\alpha_i} + [v]^{\alpha_i}, [u]^\alpha + [v]^\alpha) \\ &\leq d_H([u]^{\alpha_i}, [u]^\alpha) + d_H([v]^{\alpha_i}, [v]^\alpha) \end{aligned}$$

$$\to 0 \quad as \quad n \to \infty$$

and

$$d_H([cu]^{\alpha_i}, [cu]^\alpha) = |c| d_H([u]^{\alpha_i}, [u]^\alpha)$$

$$\to 0 \quad as \quad n \to \infty,$$

so both families of subsets also satisfy property (1.5.3). Hence, by Theorem 1.5.1, $u + v \in E^n$ and $cu \in E^n$.

In Section 1.2, the Zadeh extension principle was used to define the addition and scalar multiplication of fuzzy sets. That is,

$$(u \widetilde{+} v)(z) = \sup_{z = x + y} \min\{u(x), v(y)\} \tag{1.5.8}$$

and

$$(\tilde{c}u)(x) = u(x/c). \tag{1.5.9}$$

In E^n, these are equivalent to the level set definitions (1.5.6) and (1.5.7) respectively.

Proposition 1.5.3. *If $u, v \in E^n$ and $c \in \mathcal{R}\backslash\{0\}$, then*

$$\widetilde{u + v} = u + v \quad and \quad \widetilde{cu} = cu.$$

Proof.   Let $\alpha \in (0, 1]$. Then

$$
\begin{aligned}
\{x : \widetilde{cu}(x) \geq \alpha\} &= \{x : u(x/c) \geq \alpha\} \\
&= \{c\tilde{x} : u(\tilde{x}) \geq \alpha\} \\
&= c\{\tilde{x} : u(\tilde{x}) \geq \alpha\} = c[u]^\alpha = [cu]^\alpha
\end{aligned}
$$

and so definitions (1.5.7) and (1.5.9) coincide. Now suppose that $(u \tilde{+} v)(x) \geq \alpha$. By the definition of the supremum, there exist $x_k \in [u]^{\alpha(1-1/k)}$, $y_k \in [v]^{\alpha(1-1/k)}$ for $k = 1, 2, \ldots$ such that $x_k + y_k = z$ and so

$$(\tilde{u} + \tilde{v})(z) \geq \min\{u(x_k), v(y_k)\} \geq \alpha(1 - 1/k).$$

Since $[u]^{\alpha(1-1/k)} \to [u]^\alpha$, $[v]^{\alpha(1-1/k)} \to [v]^\alpha$ with respect to the Hausdorff metric d_H, by the compactness of all of these sets there exists $x_{k_j} \to \tilde{x}$ and $y_{k_j} \to \tilde{y}$. Hence, $x_{k_j} + y_{k+j} \to \tilde{x} + \tilde{y}$. But $x_{k_j} + y_{k_j} = z$ so $z = \tilde{x} + \tilde{y} \in [u]^\alpha + [v]^\alpha$ and

$$\{z : (\widetilde{u + v})(z) \geq \alpha\} \subseteq [u]^\alpha + [v]^\alpha.$$

Conversely, if $\tilde{x} \in [u]^\alpha$ and $\tilde{y} \in [v]^\alpha$, so that $u(\tilde{x}) \geq \alpha$ and $v(\tilde{y}) \geq \alpha$, then with $z = \tilde{x} + \tilde{y}$

$$(\widetilde{u + v})(z) \geq \min\{u(\tilde{x}), v(\tilde{y})\} \geq \alpha,$$

and so $[u]^\alpha + [v]^\alpha \subseteq \{z(\widetilde{u + v})(z) \geq \alpha\}$. Thus we have shown that

$$\{z : (\widetilde{u + v})(z) \geq \alpha\} = [u]^\alpha + [v]^\alpha = [u + v]^\alpha,$$

so definitions (1.5.6) and (1.5.8) coincide.

The concept of support function of a nonempty compact convex subset of $\mathcal{R}^n$, introduced in Section 1.4, can be usefully generalized to the fuzzy sets in E^n. Let $u \in E^n$ and define $s_u : I \times S^{n-1} \to \mathcal{R}^n$ by

$$s_u(\alpha, p) = s(p, [u]^\alpha) = \sup\{\langle p, a \rangle : a \in [u]^\alpha\} \tag{1.5.10}$$

for $(\alpha, p) \in I \times S^{n-1}$, where $s(\cdot, [u]^\alpha)$ is the support function of $[u]^\alpha$. We shall call s_u the *support function* of the fuzzy set u. Note that the supremum in

(1.5.10) is actually attained since the level set $[u]^\alpha$ is compact and so can be replaced by the maximum. Moreover, for $u, v \in E^n$

$$u = v \quad \text{if and only if} \quad s_u = s_v, \tag{1.5.11}$$

since the support function on $\mathcal{K}_C^n$ uniquely characterizes the elements of $\mathcal{K}_C^n$; see Proposition 1.4.1.

The following properties follow directly from those of the support functions of the level sets.

Proposition 1.5.4. *Let $u \in E^n$. Then the support function s_u is*

(i) *uniformly bounded on $I \times S^{n-1}$;*

(ii) *Lipschitz in $p \in S^{n-1}$ uniformly on I; and*

(iii) *for each $\alpha \in I$,*

$$d_H([u]^\alpha, [v]^\alpha) = \sup\{|s_u(\alpha, p) - s_v(\alpha, p)| : p \in S^{n-1}\}. \tag{1.5.12}$$

Proof.   Using inequalities (1.4.2) and (1.4.3)

$$|s_u(\alpha, p)| = |s(p, [u]^\alpha)| \le \|[u]^\alpha\| \|p\| \le \|[u]^0\|, \tag{1.5.13}$$

since $[u]^\alpha \subseteq [u]^0$ and $\|p\| = 1$, and

$$\begin{aligned} |s_u(\alpha, p) - s_u(\alpha, q)| &= |s(p, [u]^\alpha) - s(q, [u]^\alpha)| & (1.5.14) \\ &\le \|[u]^\alpha\| \|p - q\| & (1.5.15) \\ &\le \|[u]^0\| \|p - q\|. & (1.5.16) \end{aligned}$$

Thus (1.5.1) and (1.5.2) hold. Property (1.5.3) is a restatement of (1.4.9).

In addition, we obtain the following dependence on the membership grade.

Proposition 1.5.5. *Let $u \in E^n$. Then $s_u(\cdot, p)$ is nonincreasing and left continuous in $\alpha \in I$ for each $p \in S^{n-1}$.*

Proof.   Since $[u]^\beta \subseteq [u]^\alpha$ for $0 \le \beta \le \alpha$

$$s_u(\beta, p) = s_u(p, [u]^\beta) \le s(p, [u]^\alpha) = s_u(\alpha, p)$$

by (1.4.4), so $s_u(\cdot, p)$ is increasing for each $p \in S^{n-1}$. In addition, for a nondecreasing sequence $\alpha_i \uparrow \alpha$ in I,

$$|s_u(\alpha_i, p) - s_u(\alpha, p)| \le d_H([u]^{\alpha_i}, [u]^\alpha) \to 0$$

as $i \to \infty$ by (1.5.12) (see Lemma 1.5.1).

Example 1.5.2. *Let $u \in E^1$ have level sets $[u]^\alpha = [-1+\alpha, 1-\alpha]$ for $\alpha \in I$. Then, as $S^0 = \{-1, +1\}$, the support function s_u is given by $s_u(\alpha, \pm 1) = 1 - \alpha$ for all $\alpha \in I$.*

A fuzzy set $u \in E^n$ is called a *Lipschitzian fuzzy set* if it is a Lipschitz function of its membership grade in the sense that

$$d_H([u]^\alpha, [u]^\beta) \leq K|\alpha - \beta|$$

for all $\alpha, \beta \in I$ and some fixed, finite constant K. In view of Proposition 1.5.4, this is equivalent to the support function $s_u(\cdot, p)$ being Lipschitz uniformly in $p \in S^{n-1}$.

The subset E_C^n of *convex-sendograph fuzzy sets* consists of $\mathcal{R}^n \times I$. Hence $u \in E_C^n$ if and only if $u : \mathcal{R}^n \to I$ is a concave function over its support $[u]^0$, that is if and only if

$$(u(\lambda x + (1 - \lambda)y) \geq \lambda u(x) + (1 - \lambda)u(y)$$

for all $x, y \in [u]^0$ and $\lambda \in [0, 1]$. Note that a fuzzy convex fuzzy set is not necessarily a convex-sendograph fuzzy set. We shall denote by E_{C1}^n the subset consisting of those $u \in E_C^n$ for which the uppermost level $[u]^1$ is a singleton set.

The fuzzy sets u in E^1 are often called *fuzzy numbers*. The *triangular fuzzy numbers* are those fuzzy sets in E^1 for which the sendograph is a triangle. A triangular fuzzy number $u \in E^1$ is characterized by an ordered triple $(x_l, x_c, x_r) \in \mathcal{R}^3$ with $x_l \leq x_c \leq x_r$ such that $[u]^0 = [x_l, x_r]$ and $[u]^1 = \{x_c\}$, for then

$$[u]^\alpha = [x_c - (1 - \alpha)(x_c - x_l), x_c + (1 - \alpha)(x_r - x_c)]$$

for any $\alpha \in I$. In addition

$$d_H([u]^\alpha, [u]^\beta)] = |\alpha - \beta| \max\{x_c - x_l, x_r - x_c\}$$

so all the triangular fuzzy numbers are Lipschitzian.

1.6 The Metric Space (E^n, d)

Since E^n is a space of certain functions $u : \mathcal{R}^n \to I$ an obvious candidate for a metric on E^n is the *function space metric*

$$d(u, v) = \sup\{|u(x) - v(x)| : x \in \mathcal{R}^n\} \tag{1.6.1}$$

which measures the largest difference in the membership grades of the two fuzzy sets $u, v \in E^n$ over all points x in the base space $\mathcal{R}^n$. Note that

$$d(cu, cv) = d(u, v) \tag{1.6.2}$$

for all $u, v \in E^n$ and $c \in \mathcal{R}\backslash\{0\}$, where $cu, cv \in E^n$ are interpreted as the scalar multiplication of fuzzy sets (1.5.9), rather than the usual multiplication of a function by a scalar.

Let D_H denote the Hausdorff metric on $\mathcal{R}^{n+1}$. For any $u \in E^n$, the sendograph send(u), defined by (1.5.5), is a nonempty compact convex subset of $\mathcal{R}^n \times I \subset \mathcal{R}^{n+1}$. The *sendograph metric* D_∞ on E^n is defined in terms of the Hausdorff metric on the subspace send$(E^n) = \{$send$(u) : u \in E^n\}$ of $\mathcal{K}^{n+1}$, that is

$$D_\infty(u, v) = D_H(\text{send}(u), \text{send}(v)) \tag{1.6.3}$$

for all $u, v \in E^n$. It is certainly a metric since send$(u) =$send(v) if and only if $u = v$ in E^n. From the properties of the Hausdorff metric D_H on $\mathcal{K}^{n+1}$, it follows that

$$D_\infty(u + w, v + w') \le D_\infty(u, v) + D_\infty(w, w') \tag{1.6.4}$$

for all $u, v, w, w' \in E^n$ and

$$D_\infty(u + w, v + w) = D_\infty(u, v) \tag{1.6.5}$$

for all $u, v \in E^n$ and $w \in E^n_C$, that is with send(w) convex.

The most commonly used metrics on E^n involve the Hausdorff metric distance between the level sets of the fuzzy sets. They are, in fact, function space metrics applied to functions $\phi : I \to \mathcal{R}^+$ defined by

$$\phi(\alpha) = d_H([u]^\alpha, [v]^\alpha) \tag{1.6.6}$$

for $\alpha \in I$, where $u, v \in E^n$. In view of (1.5.12) and Proposition 1.5.5, these functions are left continuous and hence measurable on I.

The *supremum metric* d on E^n is defined by

$$d(u, v) = \sup\{d_H([u]^\alpha, [v]^\alpha) : \alpha \in I\} \tag{1.6.7}$$

for all $u, v \in E^n$ and is obviously a metric on E^n. The supremum in (1.6.7) need not be attained, so cannot be replaced by the maximum.

Example 1.6.1. *Let $u, v \in E^1$ be defined on level sets by*

$$[u]^\alpha = [v]^\alpha = [0, 1] \quad \text{for} \quad 0 \le \alpha \le \frac{1}{2},$$

and

$$[u]^\alpha = \{0\}, [v]^\alpha = [0, 2(1-\alpha)] \quad \textit{for} \quad \frac{1}{2} < \alpha \le 1,$$

so

$$\phi(\alpha) = d_H([u]^\alpha, [v]^\alpha) = \begin{cases} 0 & \textit{for} \quad 0 \le \alpha \le \frac{1}{2} \\ 2(1-\alpha) & \textit{for} \quad \frac{1}{2} < \alpha \le 1. \end{cases}$$

Then $\sup\{\phi(\alpha) : \alpha \in I\} = 1$, but this is not attained.

From the properties of the Hausdorff metric listed in Propositions 1.3.2 and 1.3.3, we get

$$d(cu, cv) = |c|d(u, v)$$

$$d(u + w, v + w) = d(u, v)$$

and

$$d(u + w, v + w') \le (u, v) + d(w, w')$$

for all $c > 0$, and all $u, v, w, w' \in E^n$.

In view of the identity (1.5.12) relating the Hausdorff metric distance between level sets and the distance between their support functions, an alternative expression for d is given by

$$d(u, v) = \sup\{|s_u(\alpha, q) - s_v(\alpha, q)| : (\alpha, q) : I \times S^{n-1}\}$$

for all $u, v \in E^n$.

The spaces (E^n, D_∞) and (E^n, d) are complete metric spaces. The proof for showing (E^n, D_∞) is complete is too long and we refer to Diamond and Kloeden [24]. We shall therefore provide the proof for (E^n, d) only.

Theorem 1.6.1. *(E^n, d) is a complete metric space.*

Proof. Let $\{u_k\}$ be a Cauchy sequence in (E^n, d). Then $\{[u_k]^\alpha\}$, for each $\alpha \in I$, is a Cauchy sequence in $(\mathcal{K}_C^n, d_H)$, which is complete, so that there exists a $C_\alpha \in \mathcal{K}_C^n$ for each $\alpha \in I$ such that

$$d_H([u_k]^\alpha, C_\alpha) \to 0 \quad \text{as} \quad k \to \infty.$$

This convergence is, in fact, uniform in $\alpha \in I$. We shall show that the family $\{C_\alpha : \alpha \in I\}$ satisfies conditions (1.5.1) and (1.5.3) and so there exists a $u \in E^n$ such that $[u]^\alpha = C_\alpha$ for $\alpha \in I$. Since the $C_\alpha \in \mathcal{K}_C^n$ for $\alpha \in I$, condition (1.5.1) is obviously satisfied. Consider $0 \le \beta < \alpha \le 1$. Then

$$\begin{aligned} d_H^*(C_\alpha, C_\beta) &\le d_H^*(C_\alpha, [u_k]^\alpha) + d_H^*([u_k]^\alpha, [u_k]^\beta) + d_H^*([u_k]^\beta, C_\beta) \\ &\le d_H(C_\alpha, [u_k]^\alpha) + d_H([u_k]^\beta, C_\beta) \\ &\to 0 \quad \text{as} \quad k \to \infty. \end{aligned}$$

Here we have used d_H^* as the Hausdorff separation in $\mathcal{R}^n$, so that $d_H^*([u_k]^\alpha,$ $[u_k]^\beta) = 0$ because $[u_k]^\alpha \subseteq [u_k]^\beta$. Hence $d_H^*(C_\alpha, C_\beta) = 0$ and, since these sets are compact, $C_\alpha \subseteq C_\beta$, so that condition (1.5.2) is satisfied. Let $\{\alpha_i\}$ be a nondecreasing sequence in I with $\alpha_i \uparrow \alpha \in I$. By the result above, $C_\alpha \subseteq C_{\alpha_i}$ for $i = 1, 2, 3, \ldots$, so

$$C_\alpha \subseteq \bigcap_{i=1}^\infty C_{\alpha_i}. \tag{1.6.8}$$

Now let $x \in \cap_{i=1}^\infty C_{\alpha_i}$, so that $x \in C_{\alpha_i}$ for $i = 1, 2, 3, \ldots$. Then

$$d_H^*(\{x\}, C_\alpha) \leq d_H^*(C_{\alpha_i}, C_\alpha)$$

$$\leq d_H^*(C_{\alpha_i}, [u_k]^{\alpha_i}) + d_H^*([u_k]^{\alpha_i}, [u_k]^\alpha)$$

$$+ d_H^*([u_k]^\alpha, C_\alpha)$$

$$\to 0 \quad \text{as} \quad k \to \infty$$

since the first and third terms converge to 0 as $k \to \infty$ uniformly in α, $\alpha_i \in I$ and the second term converges to 0 as $i \to \infty$ for each $k = 1, 2, 3, \ldots$ (condition (1.5.3) of Theorem 1.5.1). Thus $x \in C_\alpha$ and $\cap_{i=1}^\infty C_{\alpha_i} \subseteq C_\alpha$. Combining this with (1.6.8) gives

$$C_\alpha = \bigcap_{i=1}^\infty C_{\alpha_i},$$

and so condition (1.5.3) of Theorem 1.5.1 is fulfilled. Applying Theorem 1.5.1, there exists a $u \in E^n$ such that $[u]^\alpha = C_\alpha$ for $\alpha \in I$. Moreover

$$d_H([u_k]^\alpha, [u]^\alpha) \leq d_H([u_k]^\alpha, [u_j]^\alpha) + d_H([u_j]^\alpha, [u]^\alpha)$$

$$\leq d(u_k, u_j) + d_H([u_j]^\alpha, [u]^\alpha)$$

$$< \epsilon + d_H([u_j]^\alpha, [u]^\alpha)$$

for all $j, k \geq N(\epsilon)$, since$\{u_k\}$ is a Cauchy sequence in (E^n, d). Taking the limit as $j \to \infty$ we obtain $d_H([u_k]^\alpha, [u]^\alpha) \leq \epsilon$ for all $k \geq N(\epsilon)$ uniformly in $\alpha \in I$, so that $d(u_k, u) \leq \epsilon$ for all $k \geq N(\epsilon)$. Hence $u_k \to u$ in (E^n, d), which completes the proof.

1.7 Notes and Comments

All the results presented in this preliminary chapter on fuzzy sets are adapted from the monograph of Diamond and Kloeden [24], where specific references may be found. This is the first book with a good exposition which is mathematically oriented in style. See also Diamond and Kloeden [23] There are several books on fuzzy sets and application. For example, see Weaver [113], Kauffman [49], Negoita and Ralescu [79], Zimmerman [123], Kandel [47], Kauffman and Gupta [50], Pal [86], Sakawa [104], Driankov, et al. [28], and Novak [84]. For related material see also Dubois and Prade [29], Kaleva and Seikkala [45], Nguyen [81], Mizumoto and Tanaka [74, 75], Zadeh [118, 117, 116], Kruse et. al [54], and Goetschel and Voxman [35, 36].

Chapter 2

Calculus of Fuzzy Functions

2.1 Introduction

This chapter is concerned with the necessary concepts and results related to the calculus of fuzzy set-valued mappings, which we call, for short, fuzzy-valued functions or fuzzy functions. These are essentially a family of set-valued mappings and therefore we utilize the results of set-valued mappings.

Section 2.2 deals with convergence properties of fuzzy sets. In this section we adopt, for convenience, the notation u^α instead of $[u]^\alpha$ for the level sets of $u : \mathcal{R}^n \to I$. Section 2.3 discusses the measurability of fuzzy functions. In Sections 2.4 and 2.5, we develop the necessary concepts of integral and differential calculus for fuzzy functions respectively. The definition given in Section 2.4 for the integral of fuzzy functions generalizes that of Aumann [2] for set-valued mappings. Notes and comments form the content of Section 2.6.

2.2 Convergence of Fuzzy Sets

Let (Y, ρ) be a metric space and A and B two nonempty compact subsets of Y. The *Hausdorff distance* between A and B is

$$d_H(A, B) = \inf\{\epsilon : B \subset N(A, \epsilon), A \subset N(B, \epsilon)\}, \qquad (2.2.1)$$

where

$$N(A, \epsilon) = \{y \in Y : \rho(y, A) < \epsilon\}.$$

We know that d_H is a metric in the space of nonempty compact subsets of Y. Recall that we can define d_H equivalently as

$$d_H(A, B) = \max \left\{ \sup_{a \in A} \rho(a, B), \sup_{b \in B} \rho(b, A) \right\}. \qquad (2.2.2)$$

Using the continuity of $\rho(\cdot, A)$ one easily obtains:

Lemma 2.2.1. *Let A and B be two nonempty compact subsets of a metric space (Y, ρ). Then there are an $a \in A$ and a $b \in B$ such that*

$$d_H(A, B) = \rho(a, b).$$

We also need the following result.

Lemma 2.2.2. *Assume that $\{C_n\}$ is a nondecreasing (resp. nonincreasing) sequence of compact subsets of Y. If there is a subsequence of $\{C_n\}$ converging to a compact set C with respect to the Hausdorff metric then $d_H(C_n, C) \to 0$ as $n \to \infty$.*

Proof.   Let $\{C_n\}$ be a convergent subsequence of $\{C_n\}$. Since $\{C_n\}$ is also a nondecreasing sequence, we have $C_n \subset C$ for all $i = 1, 2, \ldots$. Hence

$$d_H(C_n, C) = \sup_{x \in C} \rho(x, C_{n_i}) = \sup_{x \in C} \inf_{y \in C_{n_i}} \rho(x, y).$$

If $n \geq n_i$ then $C_n \supset C_{n_i}$ and consequently for all $x \in C$,

$$\inf_{y \in C_{n_i}} \rho(x, y) \geq \inf_{y \in C_n} \rho(x, y).$$

Hence $d_H(C_n, C) \leq d_H(C_{n_i}, C)$ for all $n \geq n_i$, which implies that the sequence $\{C_n\}$ converges to C.

The nonincreasing case is treated similarly.

In addition, we have (see Castaing and Valadier [9])

$$C = \bigcap_{n=1}^{\infty} cl \left(\bigcup_{k \geq n} C_k \right). \qquad (2.2.3)$$

We say that a sequence of sets converges *metrically* if it converges with respect to the Hausdorff metric.

In the following, we utilize the Hausdorff metric to define metrics on a certain family of fuzzy sets. More precisely, recall that a fuzzy set u on $\mathcal{R}^n$, i.e., $u : \mathcal{R}^n \to I = [0, 1]$, is called *fuzzy convex* if

$$u(\lambda x + (1 - \lambda)y) \geq \min\{u(x), u(y)\}$$

for all $x, y \in \mathcal{R}^n$, $\lambda \in I$. Denote by

$$u^\alpha = \{x \in \mathcal{R}^n : u(x) \geq \alpha\}, \quad \alpha \in (0, 1]$$

the α-*level set* of u.

Recall that E^n is the family of all fuzzy sets $u : \mathcal{R}^n \to I$ with the following properties:

(i) u is upper semicontinuous,

(ii) u is fuzzy convex,

(iii) u^1 is nonempty,

(iv) the support of u,

$$\text{supp}(u) = \text{cl}(\{x : u(x) > 0\}) = \text{cl}\left(\bigcup_{\alpha > 0} u^\alpha\right),$$

is compact.

For brevity, we adopt the notation $u^0 = \text{supp}(u)$.
We immediately obtain

Lemma 2.2.3. *If $u \in E^n$, then u^α is convex compact and nonempty for all $\alpha \in I$.*

Let $u \in E^n$ and let d be the product metric on $\mathcal{R}^n \times I$ defined by the equation

$$d((x, r), (y, s)) = \max\{\|x - y\|, |r - s|\}.$$

The *endograph* of u, denoted $\text{end}(u)$, is the set

$$\text{end}(u) = \{(x, r) \in \mathcal{R}^n \times I : u(x) \geq r\}$$

and the *sendograph* of u is

$$\text{send}(u) = \text{end}(u) \cap (u^0 \times I).$$

Since u is upper semicontinuous and u^0 is compact then one easily sees that $\text{send}(u)$ is a compact subset of $\mathcal{R}^n \times I$.
We define a metric H on E^n by setting

$$H(u, v) = h^*(\text{send}(u), \text{send}(v)), \tag{2.2.4}$$

where $u, v \in E^n$ and h^* is the Hausdorff metric defined on $\mathcal{R}^n$. Then

$$d(u, v) = \sup_{\alpha > 0} d_H(u^\alpha, v^\alpha)$$

is also a metric on E^n. In fact, by definition (2.2.2) we have for all $\alpha \in (0, 1]$

$$d_H(u^\alpha, v^\alpha) \le \sup_{\substack{x \in u^\alpha \\ y \in v^\alpha}} \|x - y\| \le \sup_{\substack{x \in u^0 \\ y \in v^0}} \|x - y\| = M.$$

Since u^0 and v^0 are compact, then M and consequently $d(u, v)$, is finite. Furthermore, since d_H is a metric, d obviously satisfies the axioms of a metric.

Let $\{u_n\}$ be a sequence in E^n. We say that $\{u_n\}$ converges *levelwise* to $u \in E^n$ if for all $\alpha \in (0, 1]$

$$d_H(u_n^\alpha, u^\alpha) \to 0 \quad \text{as} \quad n \to \infty.$$

Finally denote

$$\mathcal{G} = \{u \in E^n : u \quad \text{is concave}\}.$$

In other words, if $u \in \mathcal{G}$ then

$$u(\lambda x + (1 - \lambda)y) \ge \lambda u(x) + (1 - \lambda)u(y)$$

for all $x, y \in u^0$, $\lambda \in I$.

If $u \in \mathcal{G}$, then by Lemma 2.2.3, u^0 is convex and from the concavity of u it follows that also $\text{send}(u)$ is convex.

Theorem 2.2.1. *The following implications for convergences in E^n hold true:*

(1) *Convergence in (E^n, d) implies levelwise convergence.*

(2) *Convergence in (E^n, d) implies convergence in (E^n, H).*

Proof.   The implication (1) is trivial.

For the proof of (2), let $\epsilon > 0$ be arbitrary and let $u, v \in E^n$ such that $d(u, v) < \epsilon$. Choose any $(x, \alpha) \in \text{send}(u)$.

If $\alpha > 0$ then $x \in u^\alpha$. Since $d_H(u^\alpha, v^\alpha) < \epsilon$ there exists a $y \in v^\alpha$ such that $\|x - y\| < \epsilon$. This proves that $(x, \alpha) \in N(\text{send}(v), \epsilon)$ for $(y, \alpha) \in \text{send}(v)$.

If, on the other hand, $\alpha = 0$ then $x \in u^0$. We show that $d_H(u^0, v^0) < \epsilon$. Hence we can find a $y \in v^0$ such that $\|x - y\| < \epsilon$ and as before we conclude that $(x, \alpha) \in N(\text{send}(v), \epsilon)$.

Let $\{\alpha_n\}$ be a decreasing sequence of real numbers converging to zero. Then $\{u^{\alpha_n}\}$ is a nondecreasing sequence of subsets of a compact set u^0. Hence it has a metrically convergent subsequence and thus by Lemma 2.2.2, it also converges. By (2.2.3) the limit can be expressed as

$$\bigcap_{n=1}^{\infty} cl \left(\bigcup_{m \geq n} u^{\alpha_m} \right).$$

Since $\{\alpha_n\}$ converges to zero and $\{u^{\alpha_n}\}$ is nondecreasing, we see that the limit equals u^0. So we have

$$\lim_{n \to \infty} d_H(u^{\alpha_n}, u^0) = 0 \qquad (2.2.5)$$

and similarly for v.

Since d_H is a metric we have

$$d_H(u^0, v^0) \leq d_H(u^0, u^{\alpha_n}) + d_H(u^{\alpha_n}, v^{\alpha_n}) + d_H(v^{\alpha_n}, v^0).$$

Passing to the limit yields

$$d_H(u^0, v^0) \leq \limsup_{\alpha \to 0+} d_H(u^\alpha, v^\alpha) \leq \sup_{\alpha > 0} d_H(u^\alpha, v^\alpha) = d(u, v) < \epsilon.$$

The preceding argument shows that $\text{send}(u) \subset N(\text{send}(v), \epsilon)$. Similarly, we can prove that $\text{send}(v) \subset N(\text{send}(u), \epsilon)$ and consequently

$$H(\text{send}(u), \text{send}(v)) < \epsilon.$$

It follows that the identity mapping

$$i : (E^n, d) \to (E^n, H)$$

is continuous and hence convergence in (E^n, d) implies convergence in $E^n, H)$.

We next show by examples that the implications (1) and (2) cannot be reversed.

Example 2.2.1. *Define*

$$u(x) = \begin{cases} 1 & \text{if } 0 \leq x \leq 1, \\ 0 & \text{elsewhere} \end{cases}$$

and

$$u_n(x) = \begin{cases} 1 + \frac{x-1}{n} & \text{if } 0 \leq x \leq 1, \\ 0 & \text{elsewhere,} \end{cases}$$

where $n = 1, 2, \ldots$. *Then clearly,* $\lim_{n \to \infty} H(\text{send}(u_n), \text{send}(u)) = 0$. *But* $d_H(u_n^1, u^1) = 1$ *for all* $n = 1, 2, \ldots$, *so that* $\{u_n\}$ *does not converge levelwise and hence neither in* (E^n, d).

Example 2.2.2. *Let*

$$v(x) = \begin{cases} 1 & \text{if } x = 0, \\ 0 & \text{elsewhere} \end{cases}$$

and

$$v_n(x) = \begin{cases} (1 - x)^n & \text{if } 0 \leq x \leq 1, \\ 0 & \text{elsewhere,} \end{cases}$$

where $n = 1, 2, \ldots$. *Then* $v_n^\alpha = [0, 1 - \alpha^{1/n}]$ *and*

$$d_H(v^\alpha, v_n^\alpha) = 1 - \alpha^{1/n}, \quad \alpha \in (0, 1].$$

It follows that for all $\alpha \in (0, 1]$, $d_H(v^\alpha, v_n^\alpha) \to 0$ *as* $n \to \infty$. *However,*

$$\sup_{\alpha > 0} d_H(v^\alpha, v_n^\alpha) = 1 \quad \text{for all} \quad n = 1, 2, \ldots.$$

Hence $\{v_n\}$ *does not converge in* (E^n, d). *We also see that* $d_H(v^0, v_n^0) = 1$ *for all* $n = 1, 2, \ldots$ *and hence by a theorem of Kloeden [52]* $\{v_n\}$ *does not converge in* (E^n, H).

However if we confine to $\mathcal{G}$ then the implication (1) of Theorem 2.2.1 can be reversed. Before proving that, we demonstrate some auxiliary lemmas.

Lemma 2.2.4. *Let* $\{u_n\}$ *be a sequence in* $\mathcal{G}$ *converging levelwise to* $u \in \mathcal{G}$. *Then*

$$\lim_{n \to \infty} d_H(u^0, u_n^0) = 0.$$

Proof.   From (2.2.5) we deduce that

$$\lim_{\alpha \to 0+} d_H(u^\alpha, u^0) = 0. \tag{2.2.6}$$

Let $\epsilon > 0$ be arbitrary and by (2.2.6) choose an $\alpha > 0$ such that

$$d_H(u^\alpha, u^0) < \frac{1}{2}\epsilon.$$

For this α choose an n_0 such that for all $n \geq n_0$ we have

$$d_H(u_n^\alpha, u^\alpha) < \frac{1}{2}\epsilon.$$

If now $n \geq n_0$ then

$$d_H(u^0, u_n^\alpha) \leq d_H(u^0, u^\alpha) + d_H(u^\alpha, u_n^\alpha) < \epsilon,$$

which implies that

$$u^0 \subset N(u_n^\alpha, \epsilon) \subset N(u_n^0, \epsilon). \qquad (2.2.7)$$

Conversely we prove that there exists an n_1 such that if $n \geq n_1$ then

$$u_n^0 \subset N(u^0, \epsilon). \qquad (2.2.8)$$

On the contrary, suppose that

$$u_n^0 \subset N(u^0, \epsilon) \qquad (2.2.9)$$

for infinitely many indices n. Now by the assumption, u_n^0 intersects $N(u^0, \epsilon)$ for all sufficiently large indices n. Then, by taking a subsequence if necessary, we may assume that (2.2.9) holds true for all $n = 1, 2, \ldots$ and, since u^0 is compact, there exists a sequence $\{x_n\}$ such that $x_n \in u_n^0$ and $d(x_n, u^0) = \epsilon$ for all $n = 1, 2, \ldots$.

Now choose a point $y \in u^1$. Since $\{u_n^1\}$ converges metrically to u^1 then there is a sequence $\{y_n\}$ converging to y such that $y_n \in u_n^1 \subset u_n^0$ for all $n = 1, 2, \ldots$.

Since $\{y_n\}$ converges to y and $\|y - x_n\| \geq \epsilon$ for $n \geq 1$, then there exist a sequence $\{t_n\}$ in I and an integer n_2 such that

$$\|y_n - y\| < \frac{1}{2}\epsilon \quad \text{and} \quad \|z_n - x_n\| = \frac{1}{2}\epsilon$$

whenever $n \geq n_2$, where $z_n = x_n + t_n(y_n - x_n)$.

Now for $n \geq n_2$ we have

$$\frac{1}{2}\epsilon = t_n\|y_n - x_n\| \leq t_n(\|y_n - y\| + \|y - x_n\|) < t_n(\frac{3}{2}\epsilon + \rho),$$

where $\rho = \mathrm{diam}(u^0) < \infty$, and consequently,

$$t_n > \frac{\epsilon}{3\epsilon + 2\rho} = \beta > 0.$$

But for $n \geq n_2$,

$$\rho(z_n, u^\beta) \geq \rho(z_n, u^0) \geq \rho(x_n, u^0) - \|x_n - z_n\| = \frac{1}{2}\epsilon$$

and according to the concavity of u_n,

$$u_n(z_n) \geq t_n u_n(y_n) + (1 - t_n)u_n(x_n) \geq t_n > \beta.$$

But this contradicts the fact that $\{u_n^\beta\}$ converges metrically to u^β.

Hence (2.2.8) holds true and the lemma follows from (2.2.7) and (2.2.8).

Lemma 2.2.5. *Let $u \in \mathcal{G}$ and $\alpha \in I$ be fixed. Then the function $g(\beta) = d_H(u^\beta, u^\alpha)$ is continuous at α.*

Proof.   Recall that we have already proved the continuity of g at $\alpha = 0$ (see (2.2.6)).

Let $\alpha > 0$ and choose a nonincreasing sequence $\{\alpha_k\}$ converging to α. Then $\{u^{\alpha_k}\}$ is a nondecreasing sequence with $u^{\alpha_k} \subset u^0$ for all $k = 1, 2, \ldots$. Since u^0 is compact, the sequence $\{u^{\alpha_k}\}$ has a metrically convergent subsequence and hence by Lemma 2.2.2 it converges to the limit

$$B = \bigcap_{n=1}^\infty cl \left(\bigcup_{k \geq n} u^{\alpha_k} \right).$$

Since $u^{\alpha_k} \subset u^\alpha$ for all $k = 1, 2, \ldots$ and u^α is closed, we have $B \subset u^\alpha$.

Conversely suppose that $x \in u^\alpha \backslash B$. If $u(x) > \alpha$, then $x \in u^{\alpha_k}$ for all k sufficiently large and consequently $x \in B$, which is impossible. Thus $u(y) = \alpha$ for all $y \in u^\alpha \backslash B$.

Since $x \in u^\alpha \backslash B$ and B is closed, then $M = \rho(x, B) > 0$. Choose $y \in B$ such that $u(y) > \alpha$, which is possible since $u^1 \subset B$. By the convexity of u^α we can choose a $t \in (0, 1)$ such that $z = tx + (1 - t)y \in u^\alpha$ and $\|z - x\| = (1 - t)\|x - y\| = \frac{1}{2}M$. Hence $z \in u^\alpha \backslash B$ and by the preceding paragraph $u(z) = \alpha$. Applying the concavity of u we obtain

$$\alpha = u(z) \geq tu(x) + (1 - t)u(y) > t\alpha + (1 - t)\alpha = \alpha,$$

which is impossible. So $B = u^\alpha$, which proves that g is right continuous at α.

On the other hand, let $\{\alpha_k\}$ be a nondecreasing sequence converging to α. As before, we see that $\{u^{\alpha_k}\}$ converges metrically to B. But now $u^{\alpha_k} \subset u^{\alpha_n}$ for all $k \geq n$ and hence $cl(\bigcup_{k \geq n} u^{\alpha_k}) = u^{\alpha_n}$ and consequently $B = \bigcap_{n=1}^\infty u^{\alpha_n}$. Thus by the representation theorem 1.5.1, B equals u^α.

This proves the left continuity of g and the lemma is proved.

As a corollary we have:

Corollary 2.2.1. *The function $g(\beta) = d_H(u^\beta, u^\alpha)$ is continuous.*

Proof. The triangle inequality for d_H yields

$$d_H(u^\beta, u^\alpha) - d_H(u^\beta, u^\gamma) \leq d_H(u^\gamma, u^\alpha) \leq d_H(u^\gamma, u^\beta) + d_H(u^\beta, u^\alpha).$$

The desired result follows by Lemma 2.2.5 if we let γ approach β.

Lemma 2.2.6. *Let $\{\alpha_n\}$ be a sequence of real numbers converging to $\alpha \in I$. Then under the assumptions of Lemma 2.2.4*

$$\lim_{n \to \infty} d_H(u^\alpha, u_n^{\alpha_n}) = 0.$$

Proof. Suppose on the contrary that there are an $\epsilon > 0$ and a subsequence $\{u_{n_i}^{\alpha_{n_i}}\}$ such that

$$d_H(u^\alpha, u_{n_i}^{\alpha_{n_i}}) \geq \epsilon \quad \text{for all} \quad i = 1, 2, \dots.$$

Since by Lemma 2.2.4, $\{u_n^0\}$ converges metrically to u^0 we have $u_{n_i}^{\alpha_{n_i}} \subset N(u^0, 1)$ for all i sufficiently large. Furthermore, since $\mathrm{cl}(N(u^0, 1))$ is compact it follows that $\{u_{n_i}^{\alpha_{n_i}}\}$ has a metrically convergent subsequence. Thus we may without loss of generality assume that $\{u_n^{\alpha_n}\}$ converges metrically to a compact set A with $d_H(u^\alpha, A) = M \geq \epsilon$.

We divide the rest of the proof into two cases.

Case 1. Let the sequence $\{\alpha_n\}$ be nondecreasing. By (2.2.6) and the levelwise convergence of $\{u_n\}$ to u we have

$$A = \bigcap_{n=1}^{\infty} \mathrm{cl} \left(\bigcup_{k \geq n} u_k^{\alpha_k} \right) \supset \bigcap_{n=1}^{\infty} \mathrm{cl} \left(\bigcup_{k \geq n} u_k^{\alpha} \right) = u^\alpha. \tag{2.2.10}$$

Then by Lemma 2.2.1 we can find an $x \in u^\alpha$ and a $y \in A$ such that $\|x - y\| = d_H(u^\alpha, A) = M$.

Since $\{u_n^\alpha\}$ converges metrically to u^α and $\{u_n^{\alpha_n}\}$ to A for all $n \geq 1$ we can choose $x_n \in u_n^\alpha$ and a $y_n \in u_n^{\alpha_n}$ such that

$$\lim_{n \to \infty} x_n = x \quad \text{and} \quad \lim_{n \to \infty} y_n = y.$$

Denote $z_n = \frac{1}{2}(x_n + y_n)$. Then $\lim_{n \to \infty} z_n = z = \frac{1}{2}(x+y)$ and by (2.2.10) and the choice of x and y we have

$$\rho(z, u^\alpha) = \|x - z\| = \frac{1}{2}M$$

and by the concavity of u_n

$$u_n(z_n) \geq \frac{1}{2}(u_n(x_n) + u_n(y_n)) \geq \frac{1}{2}(\alpha + \alpha_n). \qquad (2.2.11)$$

Now choose a $\beta < \alpha$ such that $d_H(u^\beta, u^\alpha) < \frac{1}{8}M$, which is possible by Lemma 2.2.5 and let n_0 be an integer such that for all $n \geq n_0$, we have

$$\|z_n - z\| < \frac{1}{8}M \quad \text{and} \quad \alpha_n \geq \beta.$$

Since $u^\beta \subset N(u^\alpha, \frac{1}{8}M)$ and $\rho(z, u^\alpha) = \frac{1}{2}M$ then

$$\rho(z, u^\beta) \geq \rho(z, N(u^\alpha, \frac{1}{8}M)) = \frac{3}{8}M.$$

Since

$$\rho(z, u^\beta) \leq \rho(z_n, u^\beta) + \|z - z_n\|$$

combining these inequalities, we obtain

$$\frac{1}{4}M < \rho(z_n, u^\beta)$$

for all $n \geq n_0$. However by (2.2.11), $z_n \in u_n^\beta$ for $n \geq n_0$. But this is impossible since $d_H(u^\beta, u_n^\beta)$ converges to zero as $n \to \infty$. This proves Case 1.

Case 2. Assume that the sequence $\{\alpha_n\}$ is nonincreasing. As in Case 1, we can show that

$$A \subset u^\alpha. \qquad (2.2.12)$$

Since $d_H(u^\alpha, A) = M > 0$ we get $u^\alpha \backslash A \neq \emptyset$. Now let $y \in u^\alpha \backslash A$ be such that $u(y) = \beta > \alpha$. Then there is an index k such that

$$y \in u^{\alpha_k} \in \bigcap_{n=1}^\infty \text{cl} \left(\bigcup_{m \geq n} u_m^{\alpha_k} \right).$$

Since $u_m^{\alpha_k} \subset u_m^{\alpha_m}$ for all $m \geq k$ and the sequence $\{\mathrm{cl}(\cup_{m \geq n} u_m^{\alpha_m})\}$ is nonincreasing it follows that

$$y \in \bigcap_{n=1}^{\infty} \mathrm{cl}\left(\bigcup_{m \geq n} u_m^{\alpha_m}\right) = A.$$

This contradiction proves that $u(y) = \alpha$ for all $y \in u^{\alpha} \backslash A$.

Since $\{u_n^1\}$ converges metrically to u^1 we have

$$u^1 = \bigcap_{n=1}^{\infty} \mathrm{cl}\left(\bigcup_{m \geq n} u_m^1\right) \subset \bigcap_{n=1}^{\infty} \mathrm{cl}\left(\bigcup_{m \geq n} u_m^{\alpha_m}\right) = A.$$

Identical reasoning as in the proof of Lemma 2.2.5 yields a contradiction.

Thus according to Cases 1 and 2, for every $\epsilon > 0$ there exists an $n(\epsilon, \alpha)$ such that

$$d_H(u^{\alpha}, u_n^{\alpha_n}) < \epsilon$$

for all $n \geq n(\epsilon, \alpha)$, which proves the lemma.

We are now ready to prove:

Theorem 2.2.2. *Let $\{u_n\}$ be a sequence in $\mathcal{G}$ converging levelwise to $u \in \mathcal{G}$. Then $\{u_n\}$ converges to u in $(\mathcal{G}, d)$.*

Proof.   Let $\alpha \in I$ and $\{\alpha_n\}$ be a sequence in I converging to α. Define a sequence of functions $f_n : I \to [0, \infty)$ by

$$f_n(\alpha) = d_H(u^{\alpha}, u_n^{\alpha}).$$

Then by the triangle inequality, we have

$$f_n(\alpha_n) = d_H(u^{\alpha_n}, u_n^{\alpha_n}) \leq d_H(u^{\alpha_n}, u^{\alpha}) + d_H(u^{\alpha}, u_n^{\alpha_n})$$

and hence by Lemmas 2.2.5 and 2.2.6, $\lim_{n \to \infty} f_n(\alpha_n) = 0$.

So $\{f_n\}$ converges continuously to zero and hence, since I is compact, it converges uniformly. But this is equivalent to convergence in $(\mathcal{G}, d)$.

Remark 2.2.1. The functions f_n are even continuous. In fact

$$|f_n(\beta) - f_n(\alpha)| \leq d_H(u^{\beta}, u^{\alpha}) + d_H(u_n^{\beta}, u_n^{\alpha})$$

and Lemma 2.2.5 gives the desired conclusion.

Example 2.2.1 shows that even in $\mathcal{G}$ convergence in $(\mathcal{G}, H)$ does not imply levelwise convergence.

Denote

$$\mathcal{G}_1 = \{u \in \mathcal{G} : u^1 \quad \text{comprises only one point}\}.$$

Then we have:

Theorem 2.2.3. *Let $\{u_n\}$ be a sequence in $\mathcal{G}_1$. If $\{u_n\}$ converges to $u \in \mathcal{G}_1$ in $(\mathcal{G}_1, H)$ then it also converges levelwise to u.*

Proof.   Since $\{u_n\}$ converges in $(\mathcal{G}_1, H)$ by a theorem of Kloeden [52] we get that for each $\eta > 0$ there exists an integer $n(\eta)$ such that for all $0 < \alpha \leq 1$ and $n \geq n(\eta)$ we have

$$u^\alpha \subset N(u_n^{\alpha-\eta}, \eta) \tag{2.2.13}$$

and

$$u_n^\alpha \subset N(u^{\alpha-\eta}, \eta). \tag{2.2.14}$$

Choose any $\epsilon > 0$. Then by Lemma 2.2.5 there exists a δ, $0 < \delta \leq \epsilon$, such that $u^{\alpha-\delta} \subset N(u^\alpha, \epsilon)$ and consequently by (2.2.14)

$$u_n^\alpha \subset N(N(u^\alpha, \epsilon), \delta) \subset N(u^\alpha, 2\epsilon) \quad \text{for all} \quad n \geq n(\delta). \tag{2.2.15}$$

Since the 1-level sets consist only of one point we see by (2.2.15) that $d_H(u_n^1, u^1)$ converges to zero.

Now let $0 < \alpha < 1$. We show that

$$u^\alpha \subset N(u_n^\alpha, 2\epsilon) \tag{2.2.16}$$

for all sufficiently large indices n.

Suppose on the contrary that there is a subsequence $\{n_i\}$ of indices for which (2.2.16) does not hold. Then we can choose a sequence $\{x_{n_i}\}$ in u^α such that $\rho(x_{n_i}, u_{n_i}^\alpha) \geq 2\epsilon$ for $i = 1, 2, \ldots$. Since u^α is compact, $\{x_{n_i}\}$ has a convergent subsequence. We may assume that $\{x_{n_i}\}$ converges to $x \in u^\alpha$. It follows by the construction of $\{x_{n_i}\}$ and (2.2.15) that $x \notin u^1$.

Let $y \in u^1$ and consider the line segment $ty + (1-t)x$, $t \in I$. From the concavity of u it follows that there is a z on that line segment such that $\|x - z\| = \frac{1}{2}\epsilon$ and $u(z) = \beta > \alpha$. Denote $\eta = \min\{\beta - \alpha, \epsilon\}$. Then by (2.2.13)

$$u^\beta \subset N(u_n^{\beta-\eta}, \eta) \subset N(u_n^\alpha, \epsilon) \quad \text{for all} \quad n \geq n(\eta).$$

Now let i_0 be such that $n_{i_0} \geq n(\eta)$. Then for all $i \geq i_0$ we have

$$\|z - x_{n_i}\| \geq \rho(x_{n_i}, u_{n_i}^\alpha) - \rho(z, u_{n_i}^\alpha) \geq 2\epsilon - \epsilon = \epsilon.$$

But this contradicts the facts that $\|z - x\| = \frac{1}{2}\epsilon$ and $\{x_{n_i}\}$ converges to x.

Thus (2.2.16) is valid and by (2.2.15) and (2.2.16), $d_H(u_n^\alpha, u^\alpha)$ converges to zero.

Combining Theorems 2.2.1–2.2.3, we obtain:

Theorem 2.2.4. *In $\mathcal{G}_1$ levelwise convergence, d-convergence and H-convergence are equivalent.*

2.3 Measurability

We shall discuss, in this section, the measurability of fuzzy functions. Let $T \subset \mathcal{R}$ be a compact interval.

Definition 2.3.1. *We say that a mapping $F : T \to E^n$ is strongly measurable if for all $\alpha \in [0,1]$ the set-valued mapping $F_\alpha : T \to \mathcal{P}_K(\mathcal{R}^n)$ defined by*

$$F_\alpha(t) = [F(t)]^\alpha$$

is (Lebesgue) measurable, when $\mathcal{P}_K(\mathcal{R}^n)$ is endowed with the topology generated by the Hausdorff metric d_H.

Puri and Ralescu [98, 99] used a more general concept for measurability. They assumed, in our terminology, that for all $\alpha \in [0,1]$ the mapping F_α has a measurable graph, that is,

$$\{(t,x)|x \in F_\alpha(t)\} \in \mathcal{M} \times \mathcal{B}(\mathcal{R}^n),$$

where $\mathcal{M}$ denotes the σ-algebra of measurable sets and $\mathcal{B}(\mathcal{R}^n)$ the Borel sets of $\mathcal{R}^n$.

However, taking into account Theorems III-2 and III-30 in Castaing and Valadier [9], it turns out that, in the setting of this section, this definition is equivalent to strong measurability.

Lemma 2.3.1. *If F is strongly measurable, then it is measurable with respect to the topology generated by d.*

Proof. Let $\epsilon > 0$ and $u \in E^n$ be arbitrary. Then

$$T_1 = \{t|d(F(t),u) \leq \epsilon\} = \bigcap_{\alpha \in [0,1]} \{t|d_H(F_\alpha(t), [u]^\alpha) \leq \epsilon\}.$$

But for all $v \in E^n$ we have

$$\lim_{k \to \infty} d([v]^{\alpha_k}, [v]^{\alpha}) = 0,$$

whenever (α_k) is a nondecreasing sequence converging to α. Thus by the triangle inequality for the metric d_H we have

$$d_H(F_\alpha(t), [u]^\alpha) \leq \limsup d_H(F_{\alpha_k}(t), [u]^{\alpha_k}),$$

where $\alpha_k \nearrow \alpha$ and consequently

$$\{t | d_H(F_\alpha(t), [u]^\alpha) \leq \epsilon\} \supset \bigcap_{k \geq 1} \{t | d_H(F_{\alpha_k}(t), [u]^{\alpha_k}) \leq \epsilon\}.$$

Thus

$$T_1 = \bigcap_{k \geq 1} \{t | d_H(F_{\alpha_k}(t), [u]^{\alpha_k}) \leq \epsilon\},$$

where $\{\alpha_k | k = 1, 2, \ldots\}$ is any denumerable dense subset of $[0, 1]$. Hence T_1 is measurable.

Lemma 2.3.2. *If $F : T \to E^n$ is continuous with respect to the metric d then it is strongly measurable.*

Proof. Let $\epsilon > 0$ be arbitrary and $t_0 \in T$. By continuity there exists a $\delta > 0$ such that

$$d(F(t), F(t_0)) < \epsilon \quad \text{whenever} \quad |t - t_0| < \delta.$$

But by the definition of d we have $d_H(F_\alpha(t), F_\alpha(t_0)) < \epsilon$ for all $|t - t_0| < \delta$, so F_α is continuous with respect to the Hausdorff metric. Therefore $F_\alpha^{-1}(U)$ is open, and hence measurable, for each open U in $\mathcal{P}_K(\mathcal{R}^n)$.

If F maps T into E^1 then $F_\alpha(t)$ is a compact interval, i.e., $F_\alpha(t) = [\lambda^\alpha(t), \mu^\alpha(t)]$. We have the following:

Lemma 2.3.3. *Let $F : T \to E^1$ be strongly measurable and denote $F_\alpha(t) = [\lambda^\alpha(t), \mu^\alpha(t)]$ for $\alpha \in [0, 1]$. Then λ^α and μ^α are measurable.*

Proof. Let $\alpha \in [0, 1]$ be fixed. Then F_α is measurable and closed valued. Consequently it has a Castaing representation (see Castaing and Valadier [9]), i.e., there exists a sequence (g_k^α) of measurable selections such that for all $t \in T$,

$$F_\alpha(t) = \overline{\{g_k^\alpha(t) | k = 1, 2, \ldots\}}.$$

But from the definition of $F_\alpha(t)$, it follows that $\lambda^\alpha = \inf g_k^\alpha$ and $\mu^\alpha = \sup g_k^\alpha$, which proves the lemma.

2.4 Integrability

A mapping $F : T \rightarrow E^n$ is called *integrably bounded* if there exists an integrable function h such that $\|x\| \leq h(t)$ for all $x \in F_0(t)$.

Definition 2.4.1. *Let $F : T \rightarrow E^n$. The integral of F over T, denoted $\int_T F(t)dt$ or $\int_a^b F(t)dt$, is defined levelwise by the equation*

$$\left[\int_T F(t)dt \right] = \int_T F_\alpha(t)dt$$
$$= \left\{ \int_T f(t)dt \,|\, f : T \rightarrow \mathcal{R}^n \text{ is a measurable selection for } F_\alpha \right\}$$

for all $0 < \alpha \leq 1$. A strongly measurable and integrably bounded mapping $F : T \rightarrow E^n$ is said to be integrable over T if $\int_T F(t)dt \in E^n$.

Note that this definition is exactly the same as employed by Puri and Ralescu [98, 99]. In their terminology, the integral is called the expected value of a fuzzy random variable.

In the following, instead of the integrals $\int_T F(t)dt$, $\int_T f(t)dt$, etc. we will write $\int F$, $\int f$, etc. When the integral is taken over a subset $S \subset T$ we will write $\int_S F$.

The following theorem due to Puri and Ralescu [98, 99] shows that certain mappings are integrable.

Theorem 2.4.1. *If $F : T \rightarrow E^n$ is strongly measurable and integrably bounded, then F is integrable.*

For a proof, see Puri and Ralescu [98].

Remark 2.4.1. Since for all $u \in E^n$,

$$\lim_{k \to \infty} d_H([u]^0, [u]^{\alpha_k}) = 0,$$

whenever (α_k) is a nonincreasing sequence converging to zero, we have (see Castaing and Valadier [9], Theorem 2.5 and Remark)

$$\lim_{k \to \infty} d_H \left(\int F_{\alpha_k}, \int F_0 \right) = 0.$$

Since

$$d_H \left(\left[\int F \right]^0, \int F_0 \right) \leq d_H \left(\left[\int F \right]^0, \int F_{\alpha_k} \right) + d_H \left(\int F_{\alpha_k}, \int F_0 \right)$$

we conclude that $[\int F]^0 = \int F_0$.

Remark 2.4.2. If $F : T \to E^1$ is integrable then in view of Lemma 2.3.3 $\int F$ is obtained by integrating the α-level curves, that is

$$\left[\int F \right]^\alpha = \left[\int \lambda^\alpha, \int \mu^\alpha \right],$$

where $[F(t)]^\alpha = [\lambda^\alpha(t), \mu^\alpha(t)]$, $\alpha \in [0,1]$.

Remark 2.4.3. Let F be integrable, $\alpha \in [0,1]$ and let $\{\sigma_n^\alpha | n = 1, 2, \ldots\}$ be a Castaing representation for F_α. Since $\int F_\alpha$ is convex and closed and includes $\int \sigma_n^\alpha$ for all $n = 1, 2, \ldots$, clearly

$$\overline{\mathrm{co}}\{\in \sigma_n^\alpha | n = 1, 2, \ldots\} \subset \int F_\alpha,$$

where $\overline{\mathrm{co}}(A)$ denotes the closed convex hull of A.

Corollary 2.4.1. *If $F : T \to E^n$ is continuous, then it is integrable.*

Proof.   By Lemma 2.3.2, F is strongly measurable. Since F_0 is continuous, $F_0(t) \in \mathcal{P}_K(\mathcal{R}^n)$ for all $t \in T$ and T is compact, then $\bigcup_{t \in T} F_0(t)$ is compact. Thus, F is integrably bounded and the conclusion follows from Theorem 2.4.1.

We shall next consider the elementary properties of the integral.

Theorem 2.4.2. *Let $F : T \to E^n$ be integrable and $c \in T$. Then*

$$\int_a^b F = \int_a^c F + \int_c^b F.$$

Proof.   Clearly the integrability of F implies that F is integrable over any subinterval of T.

Now let $\alpha \in [0,1]$ and f be a measurable selection for F_α. Since $\int_a^b f = \int_a^c f + \int_c^b f$ then we get

$$\left[\int_a^b F \right] \subset \left[\int_a^c F \right]^\alpha + \left[\int_c^b F \right]^\alpha.$$

On the other hand, let $z = \int_a^c g_1 + \int_c^b g_2$, where g_1 is a measurable selection for F_α in $[a, c]$ and g_2 is a measurable selection for F_α in $[c, b]$. Then f, defined by

$$f(t) = \begin{cases} g_1(t) & \text{if } t \in [a, c], \\ g_2(t) & \text{if } t \in (c, b], \end{cases}$$

is a measurable selection for F_α in T and

$$\int_a^b f = \int_a^c g_1 + \int_c^b g_2 = z.$$

Hence

$$\left[\int_a^c F \right]^\alpha + \left[\int_c^b F \right]^\alpha \subset \left[\int_a^b F \right]^\alpha$$

and the theorem is proved.

Corollary 2.4.2. *If* $F : T \to E^n$ *is continuous, then* $G(t) = \int_a^t F$ *is Lipschitz continuous on* T.

Proof.   Let $s, t \in T$ and assume that $s > t$. Then by Theorem 2.4.2 and the properties of d, we obtain

$$d\left(\int_a^s F, \int_a^t F \right) = d\left(\int_t^s F, \hat{0} \right),$$

where

$$\hat{0}(t) = \begin{cases} 1 & \text{if } t = 0, \\ 0 & \text{elsewhere} \end{cases}$$

Since $\bigcup_{t \in T} F_0(t)$ is compact (see the proof of Corollary 2.4.1) then there exists an $M > 0$ such that $\|x\| \leq M$ for all $x \in F_0(t)$ and $t \in T$. But this implies that

$$d(G(s), G(t)) \leq M(s - t),$$

which was to be proved.

Theorem 2.4.3. *Let* $F, G : T \to E^n$ *be integrable and* $\lambda \in \mathcal{R}$. *Then*

(i) $\int F + G = \int F + \int G$,

(ii) $\int \lambda F = \lambda \int F$,

(iii) $d(F, G)$ *is integrable,*

(iv) $d(\int F, \int G) \leq \int d(F, G)$.

Proof.   Let $\alpha \in [0, 1]$. Since F_α and G_α are compact-convex-valued it follows from Debreu [16] that the integrals $\int F_\alpha$ and $\int G_\alpha$ are in fact Bochner integrals. Hence applying equation (2.2.3) we obtain

$$\int (F + G)_\alpha = \int (F_\alpha + G_\alpha) = \int F_\alpha + \int G_\alpha,$$

which proves (i). Similar reasoning yields (ii); instead of equation (2.2.3), use equation (2.2.4).

Recall that the Hausdorff metric can be written in the form

$$d_H(A, B) = \max \left(\sup_{x \in A} \inf_{y \in B} \|x - y\|, \sup_{x \in B} \inf_{y \in A} \|x - y\| \right). \qquad (2.4.1)$$

Now for (iii), let $\{\sigma_n^\alpha | n = 1, 2, \ldots\}$ (resp. $\{\rho_n^\alpha | n = 1, 2, \ldots\}$) be a Castaing representation for F_α (resp. G_α). Applying equation (2.4.1) we get

$$d_H(F_\alpha(t), G_\alpha(t)) = \max(\sup_{n \geq 1} \inf_{k \geq 1} \|\sigma_n^\alpha(t) - \rho_k^\alpha(t)\|, \sup_{n \geq 1} \inf_{k \geq 1} \|\rho_n^\alpha(t) - \sigma_k^\alpha(t)\|),$$

which is measurable. Thus

$$d(F(t, G(t)) = \sup_{k \geq 1} d_H(F_{\alpha_k}(t), G_{\alpha_k}(t)),$$

where $\{\alpha_k | k = 1, 2, \ldots\}$ is dense in $[0, 1]$, is measurable. Furthermore

$$d(F(t), G(t)) \leq d(F(t), \hat{0}) + d(G(t), \hat{0}) \leq h_1(t) + h_2(t),$$

where h_1 and h_2 are integrable bounds for F and G respectively. Thus Proposition 7 in Royden [102], p. 82 gives us (iii).

Finally, from Debreu [16], we deduce

$$d_H \left(\int F_\alpha, \int G_\alpha \right) \leq \int d_H(F_\alpha, G_\alpha)$$

and consequently

$$d \left(\int F, \int G \right) \leq \sup_{\alpha \in [0,1]} \int d_H(F_\alpha, G_\alpha) \leq \int \sup_{\alpha \in [0,1]} d_H(F_\alpha, G_\alpha) = \int d(F, G).$$

Theorem 2.4.4. *If $F : T \to E^n$ is integrable, then the real function*

$$(t, \alpha) \to \operatorname{diam} \left[\int_a^t F \right]^\alpha, \quad t \in T, \ \alpha \in [0, 1],$$

is nondecreasing w.r.t. t on T and nonincreasing with respect to α on $[0, 1]$.

Proof. Let $t_1, t_2 \in T$ with $t_1 < t_2$. According to Theorem 2.4.1

$$\left[\int_a^{t_2} F \right]^\alpha = \left[\int_a^{t_1} F \right]^\alpha + \left[\int_{t_1}^{t_2} F \right]^\alpha,$$

which proves the first assertion. The second is trivial.

Example 2.4.1. *Let $A \in E^n$ and define $F : [0, t] \to E^n$ by $F(s) = A$ for all $0 \le s \le t$. Then*

$$\int_0^t F = tA.$$

Clearly $tA \subset \int_0^t F$. Conversely, let $\alpha \in [0, 1]$ and choose any $\int f \in \int F_\alpha$. Then $\int f$ can be expressed as a limit of a sum

$$S_n = \sum_{i=1}^n (t_i - t_{i-1}) f(\tau_i),$$

where $\{(\tau, [t_{i-1}, t_i)) | i = 1, \dots, n\}$ is a belated partition of $[0, t)$ with measure μ_n. Since $f(\tau_i) \in [A]^\alpha$ for all $i = 1, \dots, n$ and $[A]^\alpha$ is convex it follows that $S_n \in \mu_n [A]^\alpha$ for all n. As we pass to the limit then $\mu_n \nearrow t$ and consequently $\lim_{n \to \infty} d_H(t[A]^\alpha, \mu_n [A]^\alpha) = 0$. It follows that $\int f \in t[A]^\alpha$ and hence $\int_0^t F \subset tA$.

Example 2.4.2. *Define $F : T \to E^n$ by the equation*

$$F_\alpha(t) = \prod_{i=1}^n [a_i^\alpha(t), b_i^\alpha(t)], \quad \alpha \in [0, 1],$$

where $a_i^\alpha, b_i^\alpha : T \to \mathcal{R}$ are integrable, $a_i^\alpha(t) \le b_i^\alpha(t)$ for all $t \in T$ and for each $t \in T$ $a_i^\alpha(t), b_i^\alpha(t)$ are left continuous and $a_i^\alpha(t)$ (resp. $b_i^\alpha(t)$) is nondecreasing (resp. nonincreasing) with respect to α.

By Theorem 1.5.1 $F(t) \in E^n$. Now for all $\alpha \in [0, 1]$ we have

$$\int F_\alpha = \prod_{i=1}^n \left[\int a_i^\alpha, \int b_i^\alpha \right]$$

and again by Theorem 1.5.1 we see that F is integrable. The condition (1.5.3) follows from Lebesgue's convergence theorem; (1.5.1) and (1.5.2) are trivially valid.

2.5   Differentiability

Let us recall the definition of the Hukuhara difference (H-difference) [39].

Let $x, y \in E^n$. If there exists a $z \in E^n$ such that $x = y + z$ then we call z the *H-difference* of x and y, denoted $x - y$. The following definition is due to Puri and Ralescu [99].

Definition 2.5.1. *A mapping* $F : T \to E^n$ *is differentiable at* $t_0 \in T$ *if there exists a* $F'(t_0) \in E^n$ *such that the limits*

$$\lim_{h \to 0^+} \frac{F(t_0 + h) - F(t_0)}{h} \quad and \quad \lim_{h \to 0^+} \frac{F(t_0) - F(t_0 - h)}{h}$$

exist and are equal to $F'(t_0)$.

Here the limit is taken in the metric space (E^n, d). At the end points of T, we consider only the one-sided derivatives.

Remark 2.5.1. From the definition, it directly follows that if F is differentiable then the multivalued mapping F_α is Hukuhara differentiable for all $\alpha \in [0, 1]$ and

$$DF_\alpha(t) = [F'(t)]^\alpha. \tag{2.5.1}$$

Here DF_α denotes the Hukuhara derivative of F_α.

The converse results doesn't hold, since the existence of Hukuhara differences $[x]^\alpha - [y]^\alpha$, $\alpha \in [0, 1]$, does not imply the existence of the H-difference $x - y$. However, for the converse results we have the following.

Theorem 2.5.1. *Let* $F : T \to E^n$ *satisfy the assumptions:*

(a) *for each* $t \in T$, *there exists a* $\beta > 0$ *such that the H-differences* $F(t + h) - F(t)$ *and* $F(t) - F(t - h)$ *exist for all* $0 \le h < \beta$;

(b) *the set-valued mappings* F_α, $\alpha \in [0, 1]$, *are uniformly Hukuhara differentiable with derivative* DF_α, *i.e., for each* $t \in T$ *and* $\epsilon > 0$ *there exists a* $\delta > 0$ *such that*

$$d_H((F_\alpha(t + h) - F_\alpha(t))/h, \ DF_\alpha(t)) < \epsilon \tag{2.5.2}$$

and

$$d_H((F_\alpha(t) - F_\alpha(t - h))/h, \ DF_\alpha(t)) < \epsilon \tag{2.5.3}$$

for all $0 \le h < \delta$ *and* $\alpha \in [0, 1]$.

Then F is differentiable and the derivative is given by equation (2.5.1).

Proof. Consider the family $\{DF_\alpha(t) | \alpha \in [0,1]\}$. By definition $DF_\alpha(t)$ is a compact, convex and nonempty subset of $\mathcal{R}^n$.

If $\alpha_1 \leq \alpha_2$ then by assumption (a),

$$F_{\alpha_1}(t+h) - F_{\alpha_1}(t) \supset F_{\alpha_2}(t+h) - F_{\alpha_2}(t) \quad \text{for} \quad 0 \leq h < \beta$$

and consequently

$$DF_{\alpha_1}(t) \supset DF_{\alpha_2}(t). \tag{2.5.4}$$

Let $\alpha > 0$ and let (α_k) be a nondecreasing sequence converging to α. For $\epsilon > 0$ choose $h > 0$ such that equation (2.5.2) holds true. Then the triangle inequality yields

$$d_H(DF_\alpha(t), DF_{\alpha_k}(t)) \leq 2\epsilon + \frac{1}{h}d(F_\alpha(t_h) - F_\alpha(t), F_{\alpha_k}(t+h) - F_{\alpha_k}(t)).$$

By assumption (a), the rightmost term goes to zero as $k \to \infty$ and hence

$$\lim_{k\to\infty} d_H(DF_\alpha(t), DF_{\alpha_1}(t)) = 0.$$

Now by Theorem 2.2.2 and equation (2.5.4), we have

$$DF_\alpha(t) = \bigcap_{k\geq 1} DF_{\alpha_k}(t).$$

If $\alpha = 0$, then using equation (2.2.5), we deduce as before that

$$\lim_{k\to\infty} d_H(DF_0(t), DF_{\alpha_k}(t)) = 0,$$

where (α_k) is a nonincreasing sequence converging to zero, and consequently

$$DF_0(t) = \overline{\bigcup_{k\geq 1} DF_{\alpha_k}(t)}.$$

Then from Theorem 1.5.1, it follows that there is an element $u \in E^n$ such that

$$[u]^\alpha = DF_\alpha(t) \quad \text{for} \quad \alpha \in [0,1].$$

Furthermore, let $t \in T$, $\epsilon > 0$ and $\delta > 0$ be as in (b). Then

$$d((F(t+h) - F(t))/h, u) = \sup_{0<\alpha\leq 1} d_H((F_\alpha(t+H) - F_\alpha(t))/h, DF_\alpha(t)) < \epsilon$$

for all $0 \leq h < \delta$ and similarly for $d((F(t)-F(T-h))/h, u)$. Hence $F'(t) = u$ and we have the theorem.

Theorem 2.5.2. *Let $F : T \to E^1$ be differentiable.   Denote $F_\alpha(t) = [f_\alpha(t), g_\alpha(t)]$, $\alpha \in [0, 1]$. Then f_α and g_α are differentiable and*

$$[F'(t)]^\alpha = [f'_\alpha(t), g'_\alpha(t)].$$

Proof.   Now

$$[F(t + h) - F(t)]^\alpha = [f_\alpha(t + h) - f_\alpha(t), g_\alpha(t + h) - g_\alpha(t)]$$

and similarly for $[F(t) - F(t - h)]^\alpha$. Dividing by h and passing to the limit gives the conclusion.

Theorem 2.5.3. *Let $F : T \to E^n$ be differentiable on T. If $t_1, t_2 \in T$ with $t_1 \leq t_2$ then there exists a $C \in E^n$ such that $F(t_2) = F(t_1) + C$.*

Proof.   For each $s \in [t_1, t_2]$ there exists a $\delta(s) > 0$ such that the H-differences $F(s + h) - F(s)$ and $F(s) - F(s - h)$ exist for all $0 \leq h < \delta(s)$. Then we can find a finite sequence $t_1 = s_1 < s_2 < \ldots < s_n = t_2$ such that the family $\{I_{s_i} = (s_i - \delta(s_i), s_i + \delta(s_i))|\ i = 1, \ldots, n\}$ covers $[t_1, t_2]$ and $I_{s_i} \cap I_{s_i+1} \neq \emptyset$.

Pick a $v_i \in I_{s_i} \cap I_{s_i+1}$, $i = 1, \ldots, n - 1$, such that $s_i < v_i < s_{i+1}$. Then

$$F(s_{i+1}) = F(v_i) + B_1 = F(s_i) + B_2 + B_1 = F(s_i) + C_i, \quad i = 1, \ldots, n - 1,$$

for some $B_1, B_2, C_i \in E^n$. Hence

$$F(t_2) = F(t_1) + \sum_{i=1}^{n-1} C_i = F(t_1) + C.$$

As an immediate consequence we have:

Corollary 2.5.1. *If $F : T \to E^n$ is differentiable on T then for each $\alpha \in [0, 1]$ the real function $t \to diam[F(t)]^\alpha$ is nondecreasing on T.*

Theorem 2.5.4. *If $F : T \to E^n$ is differentiable then it is continuous.*

Proof.   Let $t, t + h \in T$ with $h > 0$. Then by the property of d and the triangle inequality we have

$$d(F(t + h), F(t))   =   d(F(t + h) - F(t), \hat{0})$$

$$\leq   hd((F(t + h) - F(t))/h, F'(t)) + hd(F'(t), \hat{0}),$$

where h is so small that the H-difference $F(t_h) - F(t)$ exists. By the differentiability, the right-hand side goes to zero as $h \to 0^+$ and hence F is right continuous. The left continuity is similarly proven.

A direct consequence of the properties of d is the following result.

Theorem 2.5.5. *If $F, G : T \rightarrow E^n$ are differentiable and $\lambda \in \mathcal{R}$ then $(F + G)'(t) = F'(t) + G'(t)$ and $(\lambda F)'(t) = \lambda F'(t)$.*

Theorem 2.5.6. *Let $F : T \rightarrow E^n$ be continuous. Then for all $t \in T$ the integral $G(t) = \int_a^t F$ is differentiable and $G'(t) = F(t)$.*

Proof. Note that according to Corollary 2.4.1 F is integrable. Now for $h > 0$, Theorem 2.4.2 gives

$$G(t + h) - G(t) = \int_t^{t+h} F.$$

Let $\epsilon > 0$ be arbitrary. Then by Example 2.4.1, Theorem 2.4.3 and the continuity of F, we have

$$
\begin{aligned}
d((G(t+h) - G(t))/h, F(t)) &= \frac{1}{h} d\left(\int_t^{t+h} F(s)ds, \int_t^{t+h} F(t)ds \right) \\
&\leq \frac{1}{h} \int_t^{t+h} d(F(s), F(t))ds < \epsilon
\end{aligned}
$$

for all $h > 0$ sufficiently small. Hence $\lim_{h \rightarrow 0+} (G(t_h) - G(t))/h = F(t)$ and similarly $\lim_{h \rightarrow 0+} (G(t) - G(t - h))/h = F(t)$, which proves the theorem.

Theorem 2.5.7. *Let $F : T \rightarrow E^n$ be differentiable and assume that the derivative F' is integrable over T. Then for each $s \in T$, we have*

$$F(s) = F(a) + \int_a^s F'.$$

Proof. Let $\alpha \in [0, 1]$ be fixed. We shall prove that

$$F_\alpha(s) = F_\alpha(a) + \int_a^s DF_\alpha, \tag{2.5.5}$$

where DF_α is the Hukuhara derivative of F_α, from which the theorem follows by Remark 2.5.1.

Recall that the supporting functional $\delta(\cdot, K) : \mathcal{R}^n \rightarrow \mathcal{R}$ of $K \in \mathcal{P}_K(\mathcal{R}^n)$ is defined by

$$\delta(a, K) = \sup\{a \cdot k | k \in K\},$$

where $a \cdot k$ denotes the usual scalar product of a and k. If $K_1, K_2 \in \mathcal{P}_K(\mathcal{R}^n)$ then Theorem II-18 in Castaing and Valadier [9] gives us the equation

$$d(K_1, K_2) = \sup_{\|a\|=1} |\delta(a, K_1) - \delta(a, K_2)|. \qquad (2.5.6)$$

Now let $t, t+h \in T$ with $h > 0$ so small that the H-difference $F(t+h) - F(t)$ exists. Then by Theorem II-17 in [9] we have

$$\delta(x, F_\alpha(t+h) - F_\alpha(t)) = \delta(x, F_\alpha(t+h)) - \delta(x, F_\alpha(t))$$

for all $x \in \mathcal{R}^n$, $\alpha \in [0, 1]$ and consequently

$$\delta(x, (F_\alpha(t_h) - F_\alpha(t))/h) = (\delta(x, F_\alpha(t_h)) - \delta(x, F_\alpha(t)))/h. \qquad (2.5.7)$$

Then by the differentiability of F_α and Equations (2.5.6) and (2.5.7) we obtain that $\delta(x, F_\alpha(t))$ is right differentiable and the right derivative equals $\delta(x, DF_\alpha(t))$, where x is an arbitrary element of the surface of the unit ball S in $\mathcal{R}^n$. Applying a similar reasoning for $h < 0$ we conclude that for all $x \in S$, $\delta(x, F_\alpha(t))$ is differentiable on T and

$$\frac{d}{dt}\delta(x, F_\alpha(t)) = \delta(x, DF_\alpha(t)).$$

Since $DF_\alpha(t)$ is compact and convex it can be expressed as an intersection of all closed half-spaces containing it, i.e.,

$$DF_\alpha = \bigcap_{x \in S} H_x,$$

where $H_x = \{z \in \mathcal{R}^n | x \cdot z \le \delta(x, DF_\alpha(t))\}$. Thus $DF_\alpha(t)$ equals the derivative of the set-valued mapping F_α defined by Bradley and Datko [7]. The equality (2.5.5) now follows from the same reference.

Example 2.5.1. *Let $A \in E^n$ be fixed and $r : T \to \mathcal{R}^n$ a differentiable function. Consider a mapping $F : T \to E^n$ defined by $F(t) = \widehat{r(t)} + A$, where as usual the membership function of $\widehat{r(t)}$ equals 1 at $r(t)$ and zero elsewhere. Thus $F(T)$ is a fixed fuzzy set moving along a differentiable curve r in $\mathcal{R}^n$.*

Then clearly F is differentiable and $F'(t) = \widehat{r'(t)}$. Furthermore, as a continuous mapping, it is integrable and $\int F = \widehat{\int r} + (b - a)A$.

Theorem 2.5.7 allows us to derive a mean value theorem for fuzzy mappings.

Theorem 2.5.8. *Let $F : T \to E^n$ be continuously differentiable on T. Then*

$$d(F(b), F(a)) \leq (b - a) \sup_{t \in T} d(F'(t), \hat{0}).$$

Proof. By means of Theorems 2.5.7 and 2.4.3 we obtain

$$d(F(b), F(a)) = d\left(\int F', \hat{0}\right) \leq \int d(F', \hat{0}) \leq (b - a) \sup_{t \in T} d(F'(t), \hat{0}).$$

Rolle's theorem also holds true for a fuzzy mapping $F : T \to E^1$.

Theorem 2.5.9. *Let $F : T \to E^1$ be differentiable on T. If $F(a) = F(b)$ then there exists a $t_0 \in T$ such that $F'(t_0) = \hat{0}$.*

Proof. By Theorem 2.5.3, $F(t) = F(a) + R(t)$ for some $R(t) \in E^1$. Since $F(b) = F(a)$, $\mathrm{diam}F_\alpha(b) = \mathrm{diam}F_\alpha(a)$ for all $\alpha \in [0, 1]$ and hence by Corollary 2.5.1 $\mathrm{diam}F_\alpha(t)$ is constant on T. It follows that $\mathrm{diam}F_\alpha(t) = 0$ and consequently, $R(t) = \widehat{r(t)}$ for some $r(t) \in \mathcal{R}$. Since F is differentiable, r is also differentiable on T. Furthermore $r(a) = r(b) = 0$ and so there exists a $t_0 \in T$ such that $r'(t_0) = 0$. But by Example 2.5.1, we have that $F'(t_0) = \hat{0}$.

2.6   Notes and Comments

Section 2.2 dealing with the convergence of fuzzy sets is adapted from Kaleva [41]. See Diamond and Kloeden [24] where the analysis is carried out in locally compact metric spaces. The results containing in Sections 2.3, 2.4, and 2.5 are taken from Kaleva [42]. See Puri and Ralescu [98, 99] for a more general concept of measurability. Also Dubois and Prade [30, 31, 32] have considered general concepts of derivatives and integrals. See also Diamond and Kloeden [24] for general results, and Banks and Jacobs [3], De Blasi and Lasota [14], Heilpern [38], and Hukuhara [39]. See Kloeden [52] for convergence results in a locally compact metric space X. If a linear structure is added to X, then it is finite dimensional and hence topologically isomorphic to an $\mathcal{R}^n$. See Taylor [111]. Thus one may derive results in $\mathcal{R}^n$.

Chapter 3

Fundamental Theory

3.1 Introduction

This chapter is devoted to the basic theory of fuzzy differential equations (FDEs). We begin Section 3.2 with the existence and uniqueness result for the initial value problem (IVP) employing the contraction mapping principle. Here the idea of weighted metric is utilized with effectiveness. Since the local existence result analogous to Peano's theorem in ordinary differential equations for the IVP of FDEs is still open, we prove in Section 3.3, an existence result under the stronger assumption of boundedness of the nonlinear function involved, everywhere. We establish in Section 3.4, a variety of comparison results for the solutions of FDEs which form the essential tools for studying the fundamental theory of FDEs. The comparison discussed shows how with the minimum linear structure, one can develop the theory of differential inequalities that are important in comparison principles. Section 3.5 deals with the convergence of successive approximations of the IVP of FDEs under the general uniqueness assumption of Perron type utilizing the comparison functions that is rather instructive. Continuous dependence of solutions of FDEs relative to the initial data is considered in Section 3.6. Section 3.7 investigates the global existence of solutions of FDEs. In Section 3.8, we discuss approximate solutions and error estimates between the solutions and approximate solutions. Finally, in Section 3.9, we initiate the study of stability criteria in a simpler way, suitably defining the stability concepts in the present framework. Notes and comments are provided in Section 3.10.

3.2   Initial Value Problem

Consider the initial value problem (IVP for short) for the fuzzy differential equation

$$u' = f(t, u), \quad u(t_0), \quad t_0 \geq 0 = u_0, \quad ' = \frac{d}{dt}, \qquad (3.2.1)$$

where $f \in C[J \times E^n, E^n]$, $J = [t_0, t_0 + a]$, $a > 0$. Recall that (E^n, d) is a complete metric space. Let us first note that a mapping $u : J \to E^n$ is a solution of the IVP (3.2.1) if and only if it is continuous and satisfies the integral equation

$$u(t) = u_0 + \int_{t_0}^{t} f(s, u(s))ds, \quad t \in J. \qquad (3.2.2)$$

This assertion follows from Theorems 2.5.4, 2.5.6 and 2.5.7. We also observe that if $u(t)$ satisfies (3.2.2), then

$$\mathrm{dia}[u(t)]^\alpha \geq \mathrm{dia}[u_0]^\alpha, \quad \alpha \in [0, 1],$$

where diam means the diameter of the set involved. Consequently, in view of Corollary 2.5.1, $u(t)$ is not, in general, differentiable for $t < t_0$. This means that the integral (3.2.2) cannot be extended for $t < t_0$.

As an application of the contraction mapping principle, we shall show that if $f(t, u)$ satisfies a Lipschitz condition, then the IVP (3.2.1) possesses a unique solution on J.

Theorem 3.2.1. *Assume that $f \in C[J \times E^n, E^n]$ and satisfies the Lipschitz condition*

$$d[f(t, u), f(t, v)] \leq kd[u, v], \qquad (3.2.3)$$

for $t \in J$, $u, v \in E^n$. Then the IVP (3.2.1) has a unique solution $u(t)$ on J.

Proof.   Let $c[J, E^n]$ denote the set of all continuous functions from J to E^n. Define the weighted metric

$$H(u, v) = \sup_{J} d[u(t), v(t)]e^{-\lambda t}$$

for $u, v \in C[J, E^n]$ and $\lambda > 0$ to be chosen later. Since (E^n, d) is a complete metric space, a standard argument shows that $(C[J, E^n], H)$ is also complete.

For $u \in C[J, E^n]$, we define Tu on J by the relation

$$Tu(t) = u_0 + \int_{t_0}^{t} f(s, u(s))ds. \qquad (3.2.4)$$

Then by Corollary 2.4.2, $Tu \in C[J, E^n]$. Moreover, condition (3.2.3) and the properties of the integral in Theorem 2.4.3 yield

$$d[Tu(t), Tv(t)] = d\left[u_0 + \int_{t_0}^t f(s, u(s))ds, u_0 + \int_{t_0}^t f(s, v(s))ds\right]$$

$$= d\left[\int_{t_0}^t f(s, u(s))ds, \int_{t_0}^t f(s, v(s))ds\right]$$

$$\leq \int_{t_0}^t d[f(s, u(s)), f(s, v(s))]ds$$

$$\leq k \int_{t_0}^t d[u(s), v(s)]ds, \quad t \in J.$$

Their relation, in turn, implies that for $t \in J$,

$$e^{-\lambda t} d[Tu(t), Tv(t)] \leq k e^{-\lambda t} H[u, v] \int_{t_0}^t e^{\lambda s} ds$$

$$\leq \frac{k}{\lambda} H[u, v],$$

and therefore, choosing $\lambda = 2k$, we get

$$H[Tu, Tv] \leq \frac{1}{2} H[u, v].$$

The contraction mapping principle immediately assures that there exists a unique fixed point of T, say u^*, which implies that $u^*(t)$ is the unique solution of the IVP (3.2.1) on J.

The proof is complete.

Example 3.2.1. *Let* $A, B : J \to E^1$ *be continuous. Define* $f : J \times E^1 \to E^1$ *by* $f(t, u) = A(t)u + B(t)$, *where the multiplication in* E^1 *is given by Zadeh's extension principle. If* $[A(t)]^\alpha = [a_1^\alpha(t), a_2^\alpha(t)]$ *and* $[x]^\alpha = [x_1^\alpha, x_2^\alpha]$ *then*

$$[A(t)u]^\alpha = [\min(a_1^\alpha(t)x_1^\alpha, a_2^\alpha(t)x_1^\alpha, a_1^\alpha(t)x_2^\alpha, (t)x_2^\alpha),$$

$$\max(a_1^\alpha(t)x_1^\alpha, a_2^\alpha(t)x_1^\alpha, a_1^\alpha(t)x_2^\alpha, a_2^\alpha(t)x_2^\alpha)].$$

The functions $|a_1^\alpha|, |a_2^\alpha|$ *are bounded on* J *by a constant independent of* α.

Then a straightforward computation shows that $f(t, u)$ *satisfies the assumptions of Theorem 3.2.1 and consequently the initial value problem*

$$u' = A(t)u + B(t), \quad u(t_0) = u_0,$$

has a unique solution on J.

3.3   Existence

The local existence result analogous to Peano's theorem is not valid for fuzzy differential equations, since (E^n, d) is a metric space, which is not locally compact and hence mere continuity of f in (3.2.1) is not sufficient to guarantee local existence as in finite dimensions. However, if f is continuous and bounded, we can prove an existence result. This is precisely what we plan to do in this section.

Let us continue to consider the IVP (3.2.1). As in Section 3.2, we shall employ the metric space $C[J, E^n]$ but with the unweighted metric $H[u, v] = \sup_J d[u(t), v(t)]$, $u, v \in C[J, E^n]$. For each $u \in C[J, E^n]$, we define, as before, the mapping Tu by the relation (3.2.4). We note that a fixed point of T is also a solution of the IVP (3.2.1). We can now prove the following result.

Theorem 3.3.1. *Assume that $f \in C[J \times E^n, E^n]$ and*

$$d\left[f(t, u), \hat{0}\right] \leq M, \quad t \in J, \quad u \in E^n,$$

where $\hat{0} \in E^n$ is defined as $\hat{0}(x) = 1$ if $x = 0$, and $\hat{0}(x) = 0$ if $x \neq 0$. Then the IVP (3.2.1) has a solution $u(t)$ on J.

Proof.   Let B be a bounded set in $C[J, E^n]$. The set $TB = [Tu : u \in B]$ is totally bounded if and only if it is equicontinuous and for every $t \in J$, the set $[TB](t) = [[Tu](t) : t \in J]$ is a totally bounded subset of E^n. For any $t_1, t_2 \in J$, $t_1 \leq t_2$ and $u \in B$, we get by Theorem 2.5.8,

$$d\left[Tu(t_1), Tu(t_2)\right] \leq |t_2 - t_1| \max_J d\left[f(t, u(t)), \hat{0}\right]$$

$$\leq |t_2 - t_1| M,$$

which shows that TB is equicontinuous. For $t \in J$ fixed, we have

$$d\left[Tu(t), Tu(t_1)\right] \leq |t - t_1| M$$

for every $t_1 \in J$ and $u \in B$. Consequently, the set $\{[Tu](t) : u \in B]\}$ is totally bounded in E^n. By Ascoli's theorem, we conclude that TB is a relatively compact subset of $C[J, E^n]$.

Now consider, in the metric space $(C[J, E^n], H)$, the ball $B = [u \in C[J, E^n] : H(u, \hat{0}) \leq aM]$. Then $TB \subset B$, because for $u \in C[J, E^n]$,

$$d\left[(Tu)(t), T(u)(t_0)\right] = d\left[(Tu)(t), \hat{0}\right]$$

$$\leq |t - t_0| M \leq aM.$$

Hence, defining $\tilde{0}(t) : J \to E^n$ such that $\tilde{0}(t) = \hat{0}, t \in J$, we get

$$H\left[Tu, T\hat{0}\right] = \sup_J d\left[(Tu(t)(T\tilde{0})(t)\right] \le Ma.$$

Since T is compact, by Schauder's fixed point theorem, T has a fixed point and this fixed point is a solution of (3.2.1). The proof is complete.

As we have seen, the metric space (E^n, d) has a linear structure, but is not a linear vector space. Nonetheless, it can be embedded isomorphically as a cone in a Banach space of functions in: $I \times S^{n-1} \to R$ where S^{n-1} is the unit sphere $u : R^n$, with an embedding function $u^* = \dot{F}(w)$ defined by $u^*(\alpha, x) = \max_{a\epsilon[u]^\alpha}\langle a, x \rangle$ for all $(\alpha, x) \in I \times S^{n-1}$, where $I = [0, 1]$. Fuzzy differential equations may therefore be considered and treated as differential equations on a Banach space. Following the well-developed theory of differential equations in a Banach space, one could obtain a Peano-like theorem for fuzzy differential equations on (E^n, d) by restricting to compact mappings f or to mappings f which satisfy a contraction type assumption in terms of a measure of noncompactness. See Lakshmikantham and Leela [60] for details.

The assumption of boundedness of f assumed in Theorem 3.3.1 is very strong. Thus an existence result better than Theorem 3.3.1 is still open, since the attempts in this direction are not satisfactory.

3.4   Comparison Theorems

Using the properties of $d[u, v]$ and the integral listed above, and the known theory of differential and integral inequalities, we shall establish the following comparison principles which we need for later discussion.

Theorem 3.4.1. *Assume that $f \in C[J \times E^n, E^n]$ and for $t \in J, u, v \in E^n$,*

$$d[f(t, u), f(t, v)] \le g(t, d[u, v]), \tag{3.4.1}$$

where $g \in C[J \times R_+, R_+]$ and $g(t, w)$ is nondecreasing in w for each t. Suppose further that the maximal solution $r(t, t_0, w)$ of the scalar differential equation

$$w' = g(t, w), \quad w(t_0) = w_0 \ge 0, \tag{3.4.2}$$

exists on J. Then, if $u(t), v(t)$ are any two solutions of (3.2.1) through $(t_0, u_0), (t_0, v_0)$ respectively on J, we have

$$d[u(t), v(t)] \le r(t, t_0, w_0), \quad t \in J, \tag{3.4.3}$$

provided $d[u_0, v_0] \le w_0$.

Proof. Set $m(t) = d[u(t), v(t)]$ so that $m(t_0) = d[u_0, v_0] \le w_0$. Then, in view of the properties of the metric d, we get

$$m(t) = d\left[u_0 + \int_{t_0}^t f(s, u(s))ds, v_0 + \int_{t_0}^t f(s, v(s))ds\right]$$

$$\le d\left[u_0 + \int_{t_0}^t f(s, u(s))ds, u_0 + \int_{t_0}^t f(s, v(s)ds)\right]$$

$$+ d\left[u_0 + \int_{t_0}^t f(s, v(s))ds, v_0 + \int_{t_0}^t f(s, v(s))ds\right]$$

$$= d\left[\int_{t_0}^t f(s, u(s))ds, \int_{t_0}^t f(s, v(s))ds\right] + d[u_0, v_0].$$

Now using properties of the integrals and condition (3.4.1), we observe that

$$m(t) \le m(t_0) + \int_{t_0}^t d[f(s, u(s)), f(s, v(s))]ds$$

$$\le m(t_0) + \int_{t_0}^t g(s, d[u(s), v(s)])ds$$

$$= m(t_0) + \int_{t_0}^t g(s, m(s))ds, \quad t \in J.$$

Now applying Theorem 1.9.2 given in Lakshmikantham and Leela [61], we conclude that

$$m(t) \le r(t, t_0, w_0), \quad t \in J.$$

This establishes Theorem 3.4.1.

Remark 3.4.1. If we employ the theory of differential inequalities instead of integral inequalities, we can dispense with the monotone character of $g(t, w)$ assumed in Theorem 3.4.1. This is proved in the next comparison principle.

Theorem 3.4.2. *Let the assumptions of Theorem 3.4.1 hold except the non-decreasing property of $g(t, w)$ in w. Then the conclusion (3.4.3) is valid.*

Proof. For small $h > 0$ the H-difference of $u(t+h) - u(t), v(t+h) - v(t)$ exists, and we have for $t \in I$,

$$m(t+h) - m(t) = d[u(t+h), v(t+h)] - d[u(t), v(t)].$$

Using the triangular inequality for d, we get

$$d[u(t+h), v(t+h)] \leq d[u(t+h), u(t) + hf(t, u(t))]$$

$$+ d[u(t) + hf(t, u(t)), v(t+h)],$$

and

$$d[u(t) + hf(t, u(t)), v(t+h)] \leq d[v(t) + hf(t, v(t)), v(t+h)]$$

$$+ d[u(t) + hf(t, u(t)), v(t) + hf(t, v(t))].$$

Also, we observe that

$$d[u(t) + hf(t, u(t)), v(t) + hf(t, v(t))]$$

$$\leq d[u(t) + hf(t, u(t)), u(t) + hf(t, v(t))]$$

$$+ d[u(t) + hf(t, v(t)), v(t) + hf(t, v(t))]$$

$$= d[hf(t, u(t)), hf(t, v(t))]$$

Hence, it follows that

$$\frac{m(t+h) - m(t)}{h} \leq \frac{1}{h} d[u(t+h), u(t) + hf(t, u(t))]$$

$$+ \frac{1}{h} d[v(t) + hf(t, v(t)), v(t+h)]$$

$$+ \frac{1}{h} d[hf(t, u(t)), hf(t, v(t))]$$

and consequently, in view of the properties of d and the fact that $u(t), v(t)$

are solutions of (3.4.1), we find that

$$D^+m(t) = \limsup_{h \to 0} \frac{1}{h}[m(t+h) - m(t)]$$

$$\leq \limsup_{h \to 0} d\left[\frac{u(t+h) - u(t)}{h}, f(t, u(t))\right]$$

$$+ \limsup_{h \to 0} d\left[f(t, v(t)), \frac{v(t+h) - v(t)}{h}\right]$$

$$+ d[f(t, u(t)), f(t, v(t))].$$

Here, we have used the fact that

$$d[u(t+h), u(t) + hf(t, u(t))]$$

$$= d[u(t) + z(t), u(t) + hf(t, u(t))]$$

$$= d[z(t) + u(t), u(t) + hf(t, u(t))]$$

$$= d[z(t), hf(t, u(t)]$$

$$= d[u(t+h) - u(t), hf(t, u(t))],$$

where $z(t)$ is the H-difference of $u(t+h)$ and $u(t)$. A similar argument holds for the other expression. This implies that

$$D^+m(t) \leq g(t, d[u(t), v(t)]) = g(t, m(t)), \quad t \in J$$

and therefore, the conclusion (3.4.3) follows from Theorem 1.4.1 in Lakshmikantham and Leela [61].

The next comparison result provides an estimate under weaker assumptions.

Theorem 3.4.3. *Assume that* $f \in C[I \times E^n, E^n]$ *and*

$$\limsup_{h \to 0+} \frac{1}{h}[d[u + hf(t, u), v + hf(t, v)]] - d[u, v]$$

$$\leq g(t, d[u, v]), \quad t \in J, \quad u, v \in E^n,$$

where $g \in C[J \times R_+, R]$. *The maximal solution* $r(t, t_0, w_0)$ *of (3.4.2) exists on* J. *Then the conclusion of Theorem 3.4.1 is valid.*

Proof.   Proceeding as in the proof of Theorem 3.4.2, we see that

$$
m(t + h) - m(t)
$$
$$
= \quad d[u(t + h), v(t + h)] - d[u(t), v(t)]
$$

$$
\leq \quad d[u(t + h), u(t) + hf(t, u(t))]
$$
$$
+ \quad d[v(t) + hf(t, v(t)), v(t + h)]
$$

$$
+ \quad d[hf(t, u(t)), hf(t, v(t))] - d(u(t), v(t)).
$$

$$
D^+ m(t) \quad = \quad \limsup_{h \to 0} \frac{1}{h}[m(t + h) - m(t)]
$$

$$
\leq \quad \limsup_{h \to 0+} \frac{1}{h}[d[u(t) + hf(t, u(t)), v(t) + hf(t, v(t))]]
$$

$$
- \quad d[u(t), v(t)] + \limsup_{h \to 0+} d\left[\frac{u(t + h) - u(t)}{h}, f(t, u(t))\right]
$$

$$
+ \quad \limsup_{h \to 0+} d\left[f(t, v(t)), \frac{v(t + h) - v(t)}{h}\right]
$$

$$
\leq \quad g(t, d[u(t), v(t)]) = g(t, m(t)), \quad t \in I.
$$

The conclusion follows as before by Theorem 1.4.1 in Lakshmikantham and Leela [61] and the proof is complete.

We wish to remark that in Theorem 3.4.2, $g(t, w)$ need not be non-negative and therefore the estimate (3.4.3) would be finer than the estimates in Theorems 3.4.1 and 3.4.2.

As special cases of Theorems 3.4.1, 3.4.2, and 3.4.3, we have the following important corollaries.

Corollary 3.4.1. *Assume that* $f \in C[J \times E^n, E^n]$ *and either*

(a) $d\left[f(t, u), \hat{0}\right] \leq g\left(t, d\left[u, \hat{0}\right]\right)$ *or*

(b)

$$
\limsup_{h \to 0+} \frac{1}{h}\left[d\left[u + hf(t, u), \hat{0}\right] - d\left[u, \hat{0}\right]\right] \leq g(t, d[u, 0]),
$$

where $g \in C[J \times R_+, R]$.

Then, if $d\left[u_0, \hat{0}\right] \le w_0$, we have

$$d\left[u(t), \hat{0}\right] \le r(t, t_0, w_0), \quad t \in J,$$

where $r(t, t_0, w_0)$ is the maximal solution of (3.4.2) on J.

Corollary 3.4.2. *The function $g(t, w) = \lambda(t)w, \lambda(t) > 0$ and continuous is admissible in Theorem 3.4.1 to give*

$$m(t) \le m(t_0) + \int_{t_0}^{t} \lambda(s)m(s)ds, \quad t \in J.$$

Then the Gronwall inequality implies

$$m(t) \le m(t_0) \exp\left[\int_{t_0}^{t} \lambda(s)ds\right], \quad t \in J,$$

which shows that (3.4.3) reduces to

$$d[u(t), v(t)] \le d[u_0, v_0] \exp\left[\int_{t_0}^{t} \lambda(s)ds\right], \quad t \in J.$$

Corollary 3.4.3. *In Theorem 3.4.3, the function $g(t, w) = \lambda(t)w$, where $\lambda(t)$ is the same function as in Corollary 3.4.2, is admissible and as a result we get*

$$d[u(t), v(t)] \le d[u_0, v_0] \exp\left[-\int_{t_0}^{t} \lambda(s)ds\right], \quad t \in J.$$

If $\lambda(t) = \lambda > 0$, we find that

$$d[u(t), v(t)] \le d[u_0, v_0]e^{-\lambda(t-t_0)}, \quad t \in J.$$

If $J = [t_0, \infty)$, we see that $\lim_{t \to \infty} d[u(t), v(t)] = 0$, showing the advantage of Theorem 3.4.3.

3.5   Convergence of Successive Approximations

We shall prove an existence and uniqueness result under an assumption more general than the Lipschitz-type condition considered in Section 3.2 by the method of successive approximations.

Theorem 3.5.1. *Assume that*

(a) $f \in C[R_0, E^n]$ *where* $R_0 = [J \times B(u_0, b)]$, $B(u_0, b) = [u \epsilon E^n : d[u, u_0] \leq b]$ *and* $d[f(t, u), \hat{0}] \leq M_0$ *on* R_0;

(b) $g \in C[J \times [0, 2b], R_+]$, $0 \leq g(t, w) \leq M_1$ *on* $J \times C_0[0, 2b]$, $g(t, 0) = 0$, $g(t, w)$ *is nondecreasing in* w *for each* $t \in J$ *and* $w(t) \equiv 0$ *is the unique solution of* (3.4.2) *on* J;

(c) $d[f(t, u), f(t, v)] \leq g(t, d[u, v])$ *on* R_0.

Then the successive approximations defined by

$$u_{n+1}(t) = u_0 + \int_{t_0}^{t} f(s, u_n(s))ds, \quad n = 0, 1, 2, ..., \qquad (3.5.1)$$

exist on $[t_0, t_0 + \eta]$ *where* $\eta = \min\left[a, \frac{b}{M}\right]$, $M = \max(M_0, M_1)$ *as continuous functions and converge uniformly to the unique solution* $u(t)$ *of* (3.2.1) *on* $[t_0, t_0 + \eta]$.

Proof. We have

$$
\begin{aligned}
d[u_{n+1}(t), u_0] &= d\left[u_0 + \int_{t_0}^{t} f(s, u_n(s))ds, u_0\right] \\
&= d\left[\int_{t_0}^{t} f(s, u_n(s))ds, \hat{0}\right] \leq \int_{t_0}^{t} d\left[f(s, u_n(s)), \hat{0}\right] ds \\
&\leq M_0(t - t_0) \leq M_0 a \leq b,
\end{aligned}
$$

and consequently, the successive approximations $\{u_n(t)\}$ are well defined on $[t_0, t_0 + \eta]$.

Next we shall define the successive approximations of (3.4.2) as follows

$$w_0(t) = M(t - t_0), \quad w_{n+1}(t) = \int_{t_0}^{t} g(s, w_n(s))ds, \quad t_0 \leq t \leq t_0 + \eta, \quad (3.5.2)$$

$$n = 0, 1, ...$$

An easy induction proves that $\{w_n(t)\}$ are well defined and

$$0 \leq w_{n+1}(t) \leq w_n(t) \quad t \epsilon [t_0, t_0 + \eta]. \qquad (3.5.3)$$

Since $|w_n'(t)| \leq g(t, w_{n-1}(t)) \leq M_1$, we conclude from the Ascoli–Arzela theorem and the monotonicity of the sequence $\{w_n(t)\}$, that $\lim_{n \to \infty} w_n(t) = w(t)$ uniformly on $[t_0, t_0 + \eta]$. It is also clear that $w(t)$ satisfies (3.4.2) and hence by condition (b) $w(t) \geq 0, t_0 \leq t \leq t_0 + \eta$.

We see that

$$d[u_1(t), u_0] \leq \int_{t_0}^t d\left[f(s, u_0), \hat{0}\right] ds \leq M(t - t_0) = w_0(t).$$

Suppose that

$$d[u_k(t), u_{k-1}(t)] \leq w_{k-1}(t), \quad \text{on } [t_0, t_0 + n] \text{ for some given } k.$$

Since

$$d[u_{k+1}(t), u_k(t)] \leq \int_{t_0}^t d[f(s, u_k(s)), f(s, u_{k-1}(s))]ds,$$

using condition (c) and the monotone character of $g(t, w)$ in w, we get

$$d[u_{k+1}(t), u_k(t)] \quad \leq \quad \int_{t_0}^t g(s, d[u_k(s), u_{k-1}(s)])ds$$

$$\leq \quad \int_{t_0}^t g(s, w_{k-1}(s))ds = w_k(t).$$

Thus by induction, the estimate

$$d[u_{n+1}(t), u_n(t)] \leq w_n(t), \quad t_0 \leq t \leq t_0 + \eta, \tag{3.5.4}$$

is true for all n.

Letting $v(t) = d[u_{n+1}(t), u_n(t)], t\epsilon[t_0, t_0 + \eta]$, the proof of Theorem 3.4.2 yields, for $t \in [t_0, t_0 + \eta]$,

$$D^+v(t) \leq g(t, d[u_n(t), u_{n-1}(t)]) \leq g(t, w_{n-1}(t)).$$

Let $n \leq m$. Then we obtain

$$d[u_n'(t), u_m'(t)] \quad = \quad d[f(t, u_{n-1}(t), f(t, u_{m-1}(t)))]$$

$$\leq \quad d[f(t, u_n(t)), f(t, u_{n-1}(t))] + d[f(t, u_{n-1}(t), u_{m-1}(t))]$$

$$+ d[f(t, u_m(t)), f(t, u_{m-1}(t))]$$

$$\leq \quad g(t, w_{n-1}(t)) + g(t, w_{m-1}(t)) + g(t, d[u_n(t), u_m(t)]).$$

Setting $v(t) = d[u_n(t), u_m(t)]$, the proof of Theorem 3.4.2 shows that

$$D^+ v(t) \le d[u_n'(t), u_m'(t)] \le g(t, v(t)) + 2g(t, w_{n-1}(t)), \quad t \in I,$$

in view of the monotone nature of $g(t, w)$ in w and the fact that $w_{m-1} \le w_{n-1}$ since $n \le m$ and $w_n(t)$ is a decreasing sequence. The comparison Theorem 1.4.1 in Lakshmikantham and Leela [61] then gives

$$v(t) \le r_n(t), \quad t \in I,$$

where $r_n(t)$ is the maximal solution of

$$r_n' = g(t, r_n) + 2g(t, w_{n-1}(t)), \quad r_n(t_0) = 0, \tag{3.5.5}$$

for each n. Since as $n \to \infty, 2g(t, w_{n-1}(t)) \to 0$ uniformly on $[t_0, t_0 + \eta]$, it follows by Lemma 1.3.1 in Lakshmikantham and Leela [61] that $r_n(t) \to 0$ uniformly on $[t_0, t_0 + \eta]$. This implies from (3.5.5) and the definition of $v(t)$ that $u_n(t)$ converges uniformly to $u(t)$ and it is easy to show that $u(t)$ is a solution of (3.2.1).

To show uniqueness, let $u_0(t)$ be another solution of (3.2.1). Then setting $m(t) = d[u(t), u_0(t)]$ and noting that $m(t_0) = 0$, we get $D^+ m(t) \le g(t, m(t)), t \in J$ and $m(t) \le r(t, t_0, 0), t \in J$ by Theorem 3.4.2. By the assumptions $r(t, t_0, 0) \equiv 0$ and therefore, we obtain $u(t) = u_0, t \in J$, proving uniqueness.

3.6 Continuous Dependence

In this section, we shall consider the continuous dependence of solutions of (3.2.1) with respect to initial values.

Theorem 3.6.1. *Suppose that the assumptions of Theorem 3.5.1 hold. Also further that the solutions $w(t, t_0, w_0)$ of (3.4.2) through every point (t_0, w_0) are continuous with respect to (t_0, w_0). Then the solutions $u(t, t_0, u_0)$ of (3.2.1) are continuous relative to (t_0, u_0).*

We need the following result before we prove Theorem 3.6.1.

Lemma 3.6.1. *Let $f \in C[J \times E^n, E^n]$ and let*

$$G(t, r) = \max_{d[u, u_0] \le r} d\left[f(t, u), \hat{0}\right].$$

Assume that $r^*(t, t_0, 0)$ is the maximal solution of

$$w' = G(t, w), \quad w(t_0) = 0,$$

on J. *Let* $u(t, t_0, u_0)$ *be the solution of* (3.2.1). *Then*

$$d[u(t, t_0, u_0] \le r^*(t, t_0, 0), \quad t \in J.$$

Proof.   Define $m(t) = d[u(t, t_0, u_0), u_0]$ for $t \in J$. Then Corollary 3.4.1 shows that

$$
\begin{aligned}
D^+ m(t) &\le d\left[u'(t, t_0, u_0), \hat{0}\right] \\
&= d\left[f(t, u(t, t_0, u_0)), \hat{0}\right] \\
\max_{d[u, u_0] \le m(t)} d\left[f(t, u), \hat{0}\right] &= G(t, m(t)).
\end{aligned}
$$

This implies by Theorem 1.4.1 in Lakshmikantham and Leela [61] that

$$m(t) = d[u(t, t_0, u_0), u_0] \le r^*(t, t_0, 0), \quad t \in J,$$

proving the lemma.

Proof of Theorem 3.6.1 Let $u(t) = u(t, t_0, u_0), v(t) = v(t, t_0, v_0)$ be the two solutions of (3.2.1). Then, defining $m(t) = d[u(t), v(t)]$, we get from Theorem 3.4.2 the estimate

$$d[u(t), v(t)] \le r(t, t_0, d[u_0, v_0]), \quad t \in J.$$

Since $\lim_{u_0 \to v_0} r(t, t_0, d[u_0, v_0]) = v(t, t_0, 0)$ uniformly on J and by hypothesis $r(t, t_0, 0) \equiv 0$, we get $\lim_{u_0 \to v_0} u(t, t_0, u_0)$ uniformly and hence continuity of $u(t, t_0, u_0)$ relative to u_0 is proved.

   To prove continuity with respect to t_0, let $u(t) = u(t, t_0, u_0), v(t) = v(t, \tau_0, u_0)$ be the two solutions of (3.2.1) and let $\tau_0 \ge t_0$. As before, setting $m(t) = d[u(t), v(t)]$, and noting that $m(\tau_0) = d[u(\tau_0, u_0]$, we obtain from Lemma 3.6.1,

$$m(\tau_0) \le r^*(\tau_0, t_0, 0),$$

and consequently, by Theorem 3.4.2, we arrive at

$$m(t) \le \tilde{r}(t), \quad t \ge \tau_0,$$

where $\tilde{r}(t) = \tilde{r}(t, \tau_0, r^*(\tau_0, t_0, 0))$ is the maximal solution of (3.4.2) though $(\tau_0, r^*(\tau_0, t_0, 0))$. Since $r^*(t_0, t_0, 0) = 0$, we have

$$\lim_{\tau_0 \to t_0} \tilde{r}(t, \tau_0, r^*(\tau_0, t_0, 0)) = \tilde{r}(t, t_0, 0),$$

uniformly on J. By hypothesis $\tilde{r}(t, t_0, 0) \equiv 0$ which proves the continuity of $u(t, t_0, u_0)$ relative to t_0 and the proof of Theorem 3.6.1 is complete.

   We note that the function $g(t, w) = Lw, L > 0$, is admissible in Theorems 3.5.1 and 3.6.1.

3.7   Global Existence

We consider the fuzzy differential equation

$$u' = f(t, u), \quad u(t_0) = u_0, \tag{3.7.1}$$

where $f \in C[R_+ \times E^n, E^n]$. In this section, we shall investigate the existence of solutions for $t \geq t_0$. Assuming local existence, we shall prove the following global existence result.

Theorem 3.7.1. *Assume that $f \in C[R_+ \times E^n, E^n]$ and*

$$d\left[f(t, u), \hat{0}\right] \leq g\left(t, d\left[u, \hat{0}\right]\right), \quad t, u) \in R_+ \times E^n,$$

where $g \in C[R_+^2, R_+], g(t, w)$ is nondecreasing in w for each $t \in R_+$ and the maximal solution $r(t, t_0, w_0)$ of (3.4.2) exists on $[t_0, \infty)$. Suppose further that f is smooth enough to guarantee local existence of solutions of (3.2.1) for any $(t_0, u_0) \in R_+ \times E^n$. Then the largest interval of existence of any solution $u(t, t_0, u_0)$ of (3.7.1) such that $d\left[u_0, \hat{0}\right] \leq w_0$ is $[t_0, \infty)$.

Proof.   Let $u(t) - u(t, t_0, u_0)$ be any solution of (3.7.1) with $d\left[u_0, \hat{0}\right] \leq w_0$, which exists on $[t_0, \beta), t_0 < \beta < \infty$ and the value of β cannot be increased.

Define $m(t) = d\left[u(t), \hat{0}\right]$. Then Corollary 3.4.1 shows that

$$m(t) \leq r\left(t, t_0, d\left[u, \hat{0}\right]\right) \quad t_0 \leq t < \beta. \tag{3.7.2}$$

For any t_1, t_2 such that $t_0 < t_1 < \beta$, we have

$$\begin{aligned}
d[u(t_1), u(t_2)] &= d\left[u_0 + \int_{t_0}^{t_1} f(s, u(s))ds, u_0 + \int_{t_0}^{t_2} f(s, u(s))ds\right] \\
&= d\left[\int_{t_1}^{t_2} f(s, u(s))ds, \hat{0}\right] \\
&\leq \int_{t_1}^{t_2} d\left[f(s, u(s)), \hat{0}\right] ds \\
&\leq \int_{t_1}^{t_2} g\left(s, d\left[u(s), \hat{0}\right]\right) ds.
\end{aligned}$$

The relation (3.7.2) and the nondecreasing nature of $g(t, w)$ now yields

$$\begin{aligned}
d[u(t_1), u(t_2)] &\leq \int_{t_1}^{t_2} g(s, r(s, t_0, w_0))ds \tag{3.7.3} \\
&= r(t_2, t_0, w_0) - r(t_1, t_0, w_0).
\end{aligned}$$

Since $\lim_{t \to \beta^-} r(t, t_0, w_0)$ exists and is finite by hypothesis, taking the limit as $t_1, t_2 \to \beta^-$ and using the Cauchy criterion for convergence, it follows from (3.7.3) that $\lim_{t \to \beta^-} u(t, t_0, u_0)$ exists. We then define $u(\beta, t_0, w_0) = \lim_{t \to \beta^-} u(t, t_0, u_0)$ and consider the initial value problem

$$u' = f(t, u), \quad u(\beta) = u(\beta, t_0, u_0).$$

By the assumed local existence, we see that $u(t, t_0, u_0)$ can be continued beyond β, contradicting our assumption that β cannot be continued. Hence every solution $u(t, t_0, u_0)$ of (3.7.1) such that $d\left[u_0, \hat{0}\right] \leq w_0$ exists globally on $[t_0, \infty)$ and the proof is complete.

Remark 3.7.1. Since $r(t, t_0, w_0)$ is nondecreasing because of the fact that $g(t, w) \geq 0$, if we assume that $r(t, t_0, w_0)$ is bounded on $[t_0, \infty)$, it follows that $\lim_{t \to \infty} r(t, t_0, w_0)$ exists and is finite. This, together with (3.7.2) which now holds for $t \in [t_0, \infty)$, implies that $\lim_{t \to \infty} u(t, t_0, u_0) = y \in E^n$ exists.

3.8   Approximate Solutions

We shall obtain an error estimate between the solutions and approximate solutions of IVP (3.7.1). Let us define the notion of approximate solutions.

Definition 3.8.1. *A function $v(t) = v(t, t_0, v_0, \epsilon), \epsilon > 0$, is said to be an ϵ-approximate solution of (3.7.1) if $v \in C[R_+, E^n]$, $v(t_0, t_0, v_0, \epsilon) = v_0$ and*

$$d[v'(t), f(t, v(t))] \leq \epsilon, \quad t \geq t_0.$$

In case $\epsilon = 0$, $v(t)$ is a solution of (3.7.1).

Theorem 3.8.1. *Assume that $f \in C[R_+ \times E^n, E^n]$ and for $t \geq t_0, u, v \in E^n$,*

$$d[f(t, u), f(t, v)] \leq g(t, d[u, v]), \tag{3.8.1}$$

where $g \in C[R_+^2, R_+]$. Suppose that $r(t) = r(t, t_0, w_0, \epsilon)$ is the maximal solution of

$$w' = g(t, w) + \epsilon, \quad w(t_0) = w_0 \geq 0, \tag{3.8.2}$$

existing for $t \geq t_0$. Let $u(t) = u(t, t_0, u_0)$ be any solution of (3.7.1) and $v(t) = v(t, t_0, v_0, \epsilon)$ is an ϵ-approximate solution of (3.7.1) existing for $t \geq t_0$. Then

$$d[u(t), v(t)] \leq r(t, t_0, w_0, \epsilon), \quad t \geq t_0, \tag{3.8.3}$$

provided $d[u_0, v_0] \leq w_0$.

Proof. We proceed as in the proof of Theorem 3.4.2 with $m(t) = d[u(t), v(t)]$, until we arrive at

$$D^+m(t) \leq \limsup_{h \to 0^2} d\left[\frac{u(t+h) - u(t)}{h}, f(t, u(t))\right]$$

$$+ \limsup_{h \to 0^2} d\left[f(t, v(t)), \frac{v(t+h) - v(t)}{h}\right]$$

$$+ d[f(t, u(t)), f(t, v(t))], \quad t \geq t_0.$$

This implies using the definition of approximate solution and (3.8.2) the differential inequality

$$D^+m(t) \leq g(t, m(t)) + \epsilon, \quad t \geq t_0$$

and $m(t_0) \leq w_0$. The stated estimate follows from Theorem 1.4.1 in Lakshmikantham and Leela [61].

The following corollary provides the well-known error estimate between the solution and an ϵ-approximate solution of (3.7.1).

Corollary 3.8.1. *The function $g(t, w) = Lw, L > 0$, is admissible in Theorem 3.8.1 to yield*

$$d[u(t, t_0, u_0), v[t, t_0, u_0, \epsilon)]$$

$$\leq d[u_0, v_0]e^{L(t-t_0)} + \frac{\epsilon}{L}\left(e^{L(t-t_0)} - 1\right), \quad t \geq t_0. \tag{3.8.4}$$

Proof. Since (3.8.2) in this case reduces to

$$w' = Lw + \epsilon, \quad w(t_0) = d[u_0, v_0], \tag{3.8.5}$$

it is easy to obtain the estimate (3.8.4) by solving the linear differential equation (3.8.5).

3.9 Stability Criteria

Before we proceed further to investigate stability results of fuzzy differential equations, let us note the following fact. In view of Corollary 2.5.1, the solutions of fuzzy differential equations have, in general, the property that $\text{diam}[x(t)]^\alpha$ is nondecreasing as time increases. Hence the formulation we have been working with is not suitable to reflect the rich behavior of solutions of ordinary differential equations.

Consider the following example.

Let $a \in \mathcal{E}^1$ have level sets $[a]^\alpha = [a_1^\alpha, a_2^\alpha]$ for $\alpha \in I = [0, 1]$ and suppose that a solution $x : [0, T] \to \mathcal{E}^1$ of the fuzzy differential equation

$$\frac{dx}{dt} = ax, \qquad\qquad (*)$$

on $\mathcal{E}^1$ has level sets $[x(t)]^\alpha = [x_1^\alpha(t), x_2^\alpha(t)]$ for $\alpha \in I$ and $t \in [0, T]$.

The Hukuhara derivative $\frac{dx}{dt}(t)$ also has level sets

$$\left[\frac{dx}{dt}(t) \right]^\alpha = \left[\frac{dx_1^\alpha}{dt}(t), \frac{dx_2^\alpha}{dt}(t) \right]$$

for $a \in I$ and $t \in [0, T]$ and by the extension principle, the fuzzy set $f(x(t)) = ax(t)$ has level sets

$$[ax(t)]^\alpha = [\min\{a_1^\alpha x_1^\alpha(t), a_2^\alpha x_1^\alpha(t), a_1^\alpha x_2^\alpha(t), a_2^\alpha x_2^\alpha(t)\},$$

$$\max\{a_1^\alpha x_1^\alpha(t), a_2^\alpha x_1^\alpha(t), a_1^\alpha x_2^\alpha(t), a_2^\alpha x_2(t)\}]$$

for all $\alpha \in I$ and $t \in [0, T]$. Thus the fuzzy differential equation $(*)$ is equivalent to the coupled system of ordinary differential equations

$$\frac{dx_1^\alpha}{dt} = \min\{a_1^\alpha x_1^\alpha, a_2^\alpha x_1^\alpha, a_1^\alpha x_2^\alpha, a_2^\alpha x_2^\alpha\}$$

$$\frac{dx_2^\alpha}{dt} = \max\{a_1^\alpha x_1^\alpha, a_2^\alpha x_1^\alpha, a_1^\alpha x_2^\alpha, a_2^\alpha x_2^\alpha\} \qquad\qquad (**)$$

for $\alpha \in I$. For $a = \chi_{\{-1\}} \in \mathcal{E}^1$, the fuzzy differential equation $(*)$ becomes

$$\frac{dx}{dt} = -x$$

and the system of ordinary differential equations $(**)$ reduces to

$$\frac{dx_1^\alpha}{dt} = -x_2^\alpha, \quad \frac{dx_2^\alpha}{dt} = -x_1^\alpha$$

for $\alpha \in I$. The solution corresponding to an initial value $x_0 \in \mathcal{E}^1$ with $[x_0]^\alpha = [x_{01}^\alpha, x_{02}^\alpha]$ for $\alpha \in I$ is given by

$$x_1^\alpha(t) = \frac{1}{2}(x_{01}^\alpha - x_{02}^\alpha)e^t + \frac{1}{2}(x_{01}^\alpha + x_{02}^\alpha)e^{-t},$$

$$x_2^\alpha(t) = \frac{1}{2}(x_{02}^\alpha - x_{01}^\alpha)e^t + \frac{1}{2}(x_{01}^\alpha + x_{02}^\alpha)e^{-t},$$

for $\alpha \in I$ and all $t \geq 0$.

Thus for $x_0 = \chi_{\{c_0\}}$ the solution $x(t) = \chi_{\{c_0 e^{-t}\}} \to \chi_{\{0\}}$ as $t \to \infty$. On the other hand, when $[x_0]^\alpha = [\alpha - 1, 1 - \alpha]$ for $\alpha \in I$, the solution has level sets

$$[x(t)]^\alpha = [(\alpha - 1)e^t, (1 - \alpha)e^t] = (1 - \alpha)e^t[-1, 1]$$

for all $\alpha \in I$ and $t \geq 0$. In particular, $\mathrm{diam}[x(t)]^\alpha = 2(1 - \alpha)e^t$, and hence the solution becomes fuzzier as time increases.

This shows that the stability results considered in this section and in Chapter 4 are of limited applicability. If the stability definitions are not with respect to the zero element of E^n but relative to any given solution $\phi(t) \in E^n$, then the corresponding stability criteria are perfectly in order. Nonetheless, in order to avoid complexities in formulating such definitions and the results, we have chosen to present the usual definitions and results for convenience, fully realizing their limited usefulness in this setup.

However, in Section 5.9, we describe a new concept of stability for fuzzy differential systems which includes, as a special case, the stability results, in the sense of Lyapunov, relative to a given solution. Moreover, the results presented would also cover orbital stability as well as other new notions between Lyapunov and orbital stabilities.

We shall discuss some simple stability results. We list a few definitions concerning the stability of the trivial solution of (3.7.1) which we assume to exist.

Definition 3.9.1. *The trivial solution $u = \tilde{0}$ of (3.7.1) is said to be*

(S1) *equi-stable if, for each $\epsilon > 0$ and $t_0 \in R_+$, there exists a positive function $\delta = \delta(t_0, \epsilon)$ that is continuous in t_0 for each ϵ such that*

$$d\left[u_0, \tilde{0}\right] < \delta \quad implies \quad d\left[u(t), \tilde{0}\right] < \epsilon, \quad t \geq t_0,$$

where $u(t) = u(t, t_0, u_0)$ is the solution of (3.7.1);

(S2) *uniformly stable, if the δ in (S1) is independent of t_0;*

(S3) *quasi-equi-asymptotically stable, if for each $\epsilon > 0$ and $t_0 \epsilon R_+$, there exist positive $\delta_0 = \delta_0(t_0)$ and $T = T(t_0, \epsilon)$ such that*

$$d\left[u_0, \tilde{0}\right] < \delta_0 \quad implies \quad d\left[u(t), \tilde{0}\right] < \epsilon, \quad t \geq t_0 + T;$$

(S4) *quasi-uniformly asymptotically stable, if δ_0 and T in (S3) are independent of t_0;*

(S5) *equi-asymptotically stable, if (S1) and (S3) hold simultaneously;*

(S6) *uniformly asymptotically stable, if (S2) and (S4) hold simultaneously;*

(S7) *exponentially asymptotically stable if there exists an estimate*

$$d\left[u(t), \tilde{0}\right] \leq d\left[u_0, \tilde{0}\right] e^{-\alpha(t-t_0)}, \quad \alpha > 0, \quad t \geq t_0.$$

Corresponding to the definitions (S1)–(S7), we can define the stability notions of the trivial solution $w \equiv 0$ of the scalar differential equation

$$w' = g(t, w), \quad w(t_0) = w_0 \geq 0, \tag{3.9.1}$$

where $g \in C[R_+^2, R]$ with $g(t, 0) \equiv 0$. For example, the trivial solution of (3.9.1) is equi-stable, if given $\epsilon > 0$ and $t_0 \in R_+$, there exists a $\delta = \delta(t_0, \epsilon) > 0$ that is continuous in t_0 for each ϵ such that

$$0 \leq w_0 < \delta \quad \text{implies} \quad w(t, t_0, w_0) < \epsilon, \quad t \geq t_0,$$

where $w(t, t_0, w_0)$ is any solution of (3.9.1) existing on $t \geq t_0$.

We are now in a position to prove some simple criteria for stability.

Theorem 3.9.1. *Assume that*

(i) *$f \in C[R_+ \times s(\rho), E^n]$, $s(\rho) = \left[u \epsilon E^n : d\left[u, \tilde{0}\right] < \rho\right]$, $f\left(t, \tilde{0}\right) \equiv \tilde{0}$ and for $h > 0, t \in R_+, u \in s(\rho)$,*

$$\limsup_{h \to 0^+} \frac{1}{h} \left[d\left[u + hf(t, u), \tilde{0}\right] - d\left[u, \tilde{0}\right]\right] \leq g\left(t, d\left[u, \tilde{0}\right]\right); \tag{3.9.2}$$

(ii) *$g \in C[R_+^2, R], g(t, 0) \equiv 0$.*

Then the stability properties of the trivial solution of (3.9.1) imply the corresponding stability properties of the trivial solution of the fuzzy differential equation (3.7.1) respectively.

Proof. Let the trivial solution of (3.7.1) be equi-stable. Then, given $\epsilon > 0$ and $t_0 \in R_+$, there exists a positive $\delta = \delta(t_0, \epsilon)$ with the property

$$0 \leq w_0 < \delta \quad \text{implies} \quad w(t, t_0, w_0) < \epsilon, \quad t \geq t_0, \tag{3.9.3}$$

where $w(t, t_0, w_0)$ is any solution of (3.9.1). We claim that with these ϵ, δ, the trivial solution $u(t) = \tilde{0}(t) \equiv 0$ of (3.7.1) is equi-stable.

If this is false, there would exist a solution $u(t) = u(t, t_0, u_0)$ of (3.7.1) with $d\left[u_0, \hat{0}\right] < \delta$ and $t_1 > t_0$ such that

$$d\left[u(t_1), \tilde{0}\right] = \epsilon \quad \text{and} \quad d\left[u(t), \tilde{0}\right] \leq \epsilon < \rho, \quad t_0 \leq t \leq t_1.$$

For $[t_0, t_1]$, using condition (3.9.2), Corollary 3.4.1 yields the estimate

$$d\left[u(t), \tilde{0}\right] \leq r\left(t, t_0, d\left[u_0, \hat{0}\right]\right) < \epsilon,$$

proving the claim.

One can prove similarly the other concepts of stability and we omit the details. For example, if $g(t, w) = -\alpha w, \alpha > 0$, one gets exponential asymptotic stability, since by Corollary 3.4.1, we get

$$d\left[u(t), \tilde{0}\right] \leq d\left[u_0, \hat{0}\right] e^{-\alpha(t-t_0)}, \quad t \geq t_0.$$

3.10   Notes and Comments

The results of Section 3.2 are adapted from Kaleva [42, 43] although the proof uses the metric to simplify matters. The example is also from Kaleva [42, 43]. The existence theorem in Section 3.3 is due to Nieto [83, 82]. See also Kaleva [43, 44] and Kloeden [53] for the local existence result parallel to Peano's theorem. Unfortunately, their results are not valid as described. See Friedman et al. [33, 34] for discussion and counterexamples.

The results on various comparison results given in Section 3.4 are taken from Lakshmikantham and Mohapatra [66] which depend on the well-known theory of differential inequalities. See Lakshmikantham and Leela [61]. The general result relative to the convergence of successive approximations described in Section 3.5 is due to Lakshmikantham and Mohapatra [66], which is adapted from a similar result for differential equations in a Banach space. See Lakshmikantham and Leela [60]. The work presented in Sections 3.6 and 3.7 is taken from Lakshmikantham and Mohapatra [66], which is modeled on the corresponding results in ordinary differential equations. Refer to Lakshmikantham and Leela [61]. The results dealing with approximate solutions and the error estimate are adapted from similar results in Lakshmikantham and Leela [61]. Finally, the simple stability criteria presented in Section 3.9 are modeled on the corresponding results in Lakshmikantham and Leela [59]. See for the example, Diamond and Kloeden [24].

For allied results, see Bobylev [5, 6], Chen et al. [10], Kandel et al. [47, 48], Kwun et al. [55, 56], Nayak [77], Nayak and Nanda [78], Ouyang and Wu [85], D. Park et al. [87, 88], J.Y. Park, et al. [95]–[92], Pearson [97], Neumaier [80], Seikkala [105], Song et al. [107]–[110], Wu et al. [114, 115], Zhang et al. [119, 120], and Zhou and Yu [122]. See also, Ding [26, 27], Kandel [46] and Lakshmikantham and Mohapatra [67].

Chapter 4

Lyapunov-like Functions

4.1   Introduction

In this chapter, we investigate essentially stability theory via Lyapunov-like functions. We shall also initiate development of fuzzy differential systems utilizing generalized metric spaces.

In Section 4.2 we prove a comparison result in terms of Lyapunov-like functions which serves as a vehicle for the investigation of the stability theory of Lyapunov. Section 4.3 establishes results on stability parallel to the original theorems of Lyapunov in the present framework. We provide in Section 4.4 nonuniform stability criteria employing the method of perturbing Lyapunov functions, under much weaker assumptions. Section 4.5 considers the various boundedness notions parallel to those of stability and offer sufficient conditions for the concepts of boundedness to hold.

In Section 4.6, we embark on initiating the study of fuzzy differential systems, the consideration of which leads to generalized metric spaces, in terms of which one proves comparison results utilizing the concept of vector Lyapunov functions. Section 4.7 discusses the method of vector Lyapunov functions and stability criteria. Since in this setup, one gets a comparison differential system, the study of which is sometimes difficult, we provide certain results to reduce the study of comparison systems to a single comparison equation. In Section 4.8, we consider the linear fuzzy differential system and its perturbation and develop the variation of parameters formula. We also discuss a simple periodic boundary value problem for nonhomogeneous linear fuzzy differential systems.

4.2   Lyapunov-like Functions

Consider the fuzzy differential equation

$$u' = f(t, u), \quad u(t_0) = u_0, \tag{4.2.1}$$

where $f \in C[R_+ \times S(\rho), E^n]$ and $S(\rho) = [u \in E^n : d[u, \hat{0}] < \rho]$. We assume that $f(t, \hat{0}) = \hat{0}$ so that we have the trivial solution for (4.2.1).

To investigate stability criteria, the following comparison result in terms of a Lyapunov function is very important and can be proved via the theory of differential inequalities. Here the Lyapunov function serves as a vehicle to transform the fuzzy differential equation into a scalar comparison differential equation and therefore it is enough to consider the stability properties of the simpler comparison equation.

Theorem 4.2.1. *Assume that*

  (i) $V \in C[R_+ \times S(\rho), R_+], |V(t, u_1) - V(t, u_2)| \leq Ld[u_1, u_2], L > 0$ *and*

  (ii) $D^+V(t, u) \equiv \lim_{h \to 0+} \sup \frac{1}{h} [V(t+h, u+hf(t, u)) - V(t, u)] \leq g(t, V(t, u))$, *where* $g \in C[R_+^2, R]$.

Then, if $u(t)$ *is any solution of (4.2.1) existing on* $[t_0, \infty)$ *such that* $V(t_0, u_0) \leq w_0$, *we have*

$$V(t, u(t)) \leq r(t, t_0, w_0), \quad t \geq t_0,$$

where $r(t, t_0, w_0)$ *is the maximal solution of the scalar differential equation*

$$w' = g(t, w), \quad w(t_0) = w_0 \geq 0, \tag{4.2.2}$$

existing on $[t_0, \infty)$.

Proof.   Let $u(t)$ be any solution of (4.2.1) existing on $[t_0, \infty)$. Define $m(t) = V(t, u(t))$ so that $m(t_0) = V(t_0, u_0) \leq w_0$. Now for small $h > 0$,

$$m(t + h) - m(t) \; = \; V(t + h, u(t + h)) - V(t, u(t))$$

$$= \; V(t + h, u(t + h)) - V(t + h, u(t) + hf(t, u(t)))$$

$$+ V(t + h, u(t) + hf(t, u(t))) - V(t, u(t)),$$

$$\leq \; Ld[u(t + h), u(t) + hf(t, u(t))]$$

$$+ V(t + h, u(t) + hf(t, u(t))) - V(t, u(t)),$$

using the Lipschitz condition given in (i). Thus

$$D^+m(t) = \lim_{h \to 0^+} \sup \frac{1}{h}[m(t+h) - m(t)]$$

$$\leq D^+V(t, u(t)) + L \lim_{h \to 0^+} \sup \frac{1}{h}[d[u(t+h), u(t) + hf(t, u(t))]].$$

Let $u(t + h) = u(t) + z(t)$, where $z(t)$ is the H-difference for small $h > 0$ which is assumed to exist. Hence employing the properties of $d[u, v]$, we see that

$$d[u(t+h), u(t) + hf(t, u(t)) = d[u(t) + z(t), u(t) + hf(t, u(t))]$$

$$= d[z(t), hf(t, u(t))]$$

$$= d[u(t+h) - u(t), hf(t, u(t))].$$

Consequently

$$\frac{1}{h}d[u(t+h), u(t) + hf(t, u(t))] = d\left[\frac{u(t+h) - u(t)}{h}, f(t, u(t))\right]$$

and therefore

$$\lim_{h \to 0^+} \sup \frac{1}{h}[d[u(t+h), u(t) + hf(t, u(t))]]$$

$$= \lim_{h \to 0^+} \sup \frac{1}{h}\left[d\left[\frac{u(t+h) - u(t)}{h}, f(t, u(t))\right]\right]$$

$$= d[u'(t), f(t, u(t))] = 0,$$

since $u(t)$ is the solution of (4.2.1). We therefore have the scalar differential inequality

$$D^+m(t) \leq g(t, m(t)), \quad m(t_0) \leq w_0, \quad t \geq t_0,$$

which by the theory of differential inequalities (see Lakshmikantham and Leela [61]) implies

$$m(t) \leq r(t, t_0, w_0), \quad t \geq t_0.$$

This proves the claimed estimate of the theorem.

The following corollaries are useful.

Corollary 4.2.1. *The function $g(t, w) \equiv 0$ is admissible in Theorem 4.2.1 to yield the estimate*

$$V(t, u(t)) \leq V(t_0, u_0), \quad t \geq t_0.$$

Corollary 4.2.2. *If, in Theorem 4.2.1, we strengthen the assumption on $D^+V(t, u)$ to*

$$D^+V(t, u) \leq -C[w(t, u)] + g(t, V(t, u)),$$

where $w \in C[R_+ \times S(\rho), R_+]$, $C \in \mathcal{K} = [a \in C[[o, \rho), R_+] : a(w)$ is increasing in w and $a(0) = 0]$, and $g(t, w)$ is nondecreasing in w for each $t \in R_+$, then we get the estimate

$$V(t, u(t)) + \int_{t_0}^t C[w(s, u(s))]ds \leq r(t, t_0, w_0), \quad t \geq t_0,$$

whenever $V(t_0, u_0) \leq w_0$.

Proof.   Set $L(t, u(t)) = V(t, u(t)) + \int_{t_0}^t C[w(s, u(s))]ds$ and note that

$$
\begin{aligned}
D^+L(t, u(t)) &\leq D^+V(t, u(t)) + C[w(t, u(t))] \\
&\leq g(t, V(t, u(t))) \\
&\leq g(t, L(t, u(t))),
\end{aligned}
$$

using the monotone character of $g(t, w)$. We then get immediately by Theorem 4.2.1 the estimate

$$L(t, u(t)) \leq r(t, t_0, w_0), \quad t \geq t_0,$$

where $u(t)$ is any solution of (4.2.1). This implies the stated estimate.
    A simple example of $V(t, u)$ is $d[u, \hat{0}]$ so that

$$D^+V(t, u) = \lim_{h \to 0^+} \sup \frac{1}{h}[d[u + hf(t, u), \hat{0}] - d[u, \hat{0}]].$$

4.3   Stability Criteria

Having necessary comparison results in terms of Lyapunov-like functions, it is easy to establish stability results analogous to original Lyapunov results for fuzzy differential equations.
    Let us start with the following result on equi-stability.

Theorem 4.3.1. *Assume that the following hold:*

(i) $V \in C[R_+ \times S(\rho), R_+]$, $|V(t, u_1) - V(t, u_2)| \leq Ld[u_1, u_2], L > 0$ *and for $(t, u) \in R_+ \times S(\rho)$, where $S(\rho) = [u \in E^n : d[u, \hat{0}] < \rho]$,*

$$D^+ V(t, u) \equiv \limsup_{h \to 0^+} \frac{1}{h}[V(t + h, u + hf(t, u)) - V(t, u)] \leq 0; \quad (4.3.1)$$

(ii) $b(d[u, \hat{0}]) \leq V(t, u) \leq a(t, d[u, \hat{0}])$ *for $(t, u) \in R_+ \times S(\rho)$ where $b, a(t, \cdot)$ $\in \mathcal{K} = [\sigma \in C[(0, \rho), R_+] : \sigma(0) = 0$ and $\sigma(w)$ is increasing in $w]$.*

Then the trivial solution of (4.2.1) is equi-stable.

Proof.   Let $0 < \epsilon < \rho$ and $t_0 \in R_+$ be given. Choose a $\delta = \delta(t_0, \epsilon)$ such that

$$a(t_0, \delta) < b(\epsilon). \quad (4.3.2)$$

We claim that with this δ, equi-stability holds. If not, there would exist a solution $u(t) = u(t, t_0, u_0)$ of (4.2.1) and a $t_1 > t_0$ such that

$$d[u(t_1), \tilde{0}] = \epsilon \quad \text{and} \quad d[u(t), \tilde{0}] \leq \epsilon < \rho, \quad t_0 \leq t \leq t_1. \quad (4.3.3)$$

By Corollary 4.2.1, we then have

$$V(t, u(t)) \leq V(t_0, u_0), \quad t_0 \leq t \leq t_1.$$

Consequently, using (ii), (4.3.2) and (4.3.3), we arrive at the following contradiction

$$b(\epsilon) = b(d[u(t_1), \tilde{0}]) \leq V(t_1, u(t_1)) \leq V(t_0, u_0) \leq a(t_0, d[u_0, \hat{0}]) < b(\epsilon).$$

Hence equi-stability holds, completing the proof.

The next result provides sufficient conditions for equi-asymptotic stability. In fact, it gives exponential asymptotic stability.

Theorem 4.3.2. *Let the assumptions of Theorem 4.3.1 hold except that the estimate (4.3.1) be strengthened to*

$$D^+ V(t, u) \leq -\beta V(t, u), \quad (t, u) \in R_+ \times S(\rho). \quad (4.3.4)$$

Then the trivial solution of (4.2.1) is equi-asymptotically stable.

Proof.   Clearly the trivial solution of (4.2.1) is equi-stable. Hence taking $\epsilon = \rho$ and designating $\delta_0 = \delta(t_0, \rho)$, we have byTheorem 4.3.1,

$$d[u_0, \tilde{0}] < \delta_0 \quad \text{implies} \quad d[u(t), \tilde{0}] < \rho, \quad t \geq t_0.$$

Consequently, we get from assumption (4.3.4), the estimate

$$V(t, u(t)) \leq V(t_0, u_0) \exp[-\beta(t - t_0)], \quad t \geq t_0.$$

Given $\epsilon > 0$, we choose $T = T(t_0, \epsilon) = \frac{1}{\beta} \ln \frac{a(t_0, \delta_0)}{b(\epsilon)} + 1$. Then it is easy to see that

$$b(d[u(t), \tilde{0}]) \leq V(t, u(t)) \leq a(t_0, \delta) e^{-\beta(t-t_0)} < b(\epsilon), \quad t \geq t_0 + T.$$

The proof is complete.

We shall next consider uniform stability criteria.

Theorem 4.3.3. *Assume that, for $(t, u) \in R_+ \times S(\rho) \cap S^c(\eta)$ for each $0 < \eta < \rho$, $V \in C[R_+ \times S(\rho) \cap S^c(\eta), R_+]$, $|V(t, u_1) - V(t, u_2)| \leq Ld[u_1, u_2]$, $L > 0$*

$$D^+ V(t, u) \leq 0 \tag{4.3.5}$$

and

$$b(d[u, \tilde{0}]) \leq V(t, u) \leq a[d[u, \tilde{0}]), \quad a, b \in \mathcal{K}. \tag{4.3.6}$$

Then the trivial solution of (4.2.1) is uniformly stable.

Proof.   Let $0 < \epsilon < \rho$ and $t_0 \in R_+$ be given. Choose $\delta = \delta(\epsilon) > 0$ such that $a(\delta) < b(\epsilon)$. Then we claim that with this δ, uniform stability follows. If not, there would exist a solution $u(t)$ of (4.2.1), and a $t_2 > t_1 > t_0$ satisfying

$$d[u(t_1), \tilde{0}] = \delta, \quad d[u(t_2), \tilde{0}] = \epsilon \quad \text{and} \quad \delta \leq d[u(t), \tilde{0}] \leq \epsilon < \rho. \tag{4.3.7}$$

Taking $\eta = \delta$, we get from (4.3.5), the estimate

$$V(t_2, u(t_2)) \leq V(t_1, u(t_1)),$$

and therefore, (4.3.6) and (4.3.7) together with the definition of δ, yield

$$
\begin{aligned}
b(\epsilon) &= b(d[u(t_2), \tilde{0}) \\
&\leq V(t_2, u(t_2)) \\
&\leq V(t_1, u(t_1)) \\
&\leq a(d[u(t_1), \tilde{0}]) \\
&= a(\delta) \\
&< b(\epsilon).
\end{aligned}
$$

This contradiction proves uniform stability, completing the proof.

Finally, we shall prove uniform asymptotic stability.

Theorem 4.3.4. *Let the assumptions of Theorem 4.3.3 hold except that (4.3.5) is strengthened to*

$$D^+V(t, u) \leq -c(d[u, \hat{0}]), \quad C \in \mathcal{K}. \tag{4.3.8}$$

Then the trivial solution of (4.3.1) is uniformly asymptotically stable.

Proof. By Theorem 4.3.3, uniform stability follows and so for $\epsilon = \rho$, we designate $\delta_0 = \delta_0(\rho)$. This means that

$$d[u_0, \hat{0}] < \delta_0 \quad \text{implies} \quad d[u(t), \tilde{0}] < \rho, \quad t \geq t_0.$$

In view of uniform stability, it is enough to show that there exists a t^* such that for $t_0 \leq t^* \leq t_0 + T$, where $T = 1 + \frac{a(\delta_0)}{b(\delta)}$,

$$d[u(t^*), \tilde{0}] < \delta. \tag{4.3.9}$$

If this is not true, $\delta \leq d[u(t), \tilde{0}]$ for $t_0 \leq t \leq t_0 + T$. Then (4.3.8) gives

$$V(t, u(t)) \leq V(t_0, u_0) - \int_{t_0}^{t} C(d[u(s), \tilde{0}])ds, \quad t_0 \leq t \leq t_0 + T.$$

As a result, we have, in view of the choice of T,

$$0 \leq V(t_0 + T, u(t_0 + T)) \leq a(\delta_0) - C(\delta)T < 0,$$

a contradiction. Hence there exists a t^* satisfying (4.3.9) and uniform stability then shows that

$$d[u_0, \hat{0}] < \delta_0 \quad \text{implies} \quad d[u(t), \tilde{0}] < \epsilon, \quad t \geq t_0 + T,$$

and the proof is complete.

4.4 Nonuniform Stability Criteria

In Section 4.3, we discussed stability results parallel to Lyapunov's original theorems for fuzzy differential equations. We note that in proving nonuniform stability concepts, one needs to impose assumptions everywhere in $R_+ \times S(\rho)$, whereas to investigate uniform stability notions, it is enough to assume conditions in $R_+ \times S(\rho \cap S^c(\eta))$ for $0 < \eta < \rho$, where $S^c(\eta)$ denotes the complement of $S(\eta)$. The question therefore arises whether one can prove nonuniform stability notions under less restrictive assumptions. The answer is yes and one needs to employ the method of perturbing Lyapunov functions to achieve this. This is what we plan to do in this section.

We begin with the following result which provides nonuniform stability criteria under weaker assumptions.

Theorem 4.4.1. *Assume that*

(A1) $V_1 \in C[R_+ \times S(\rho), R_+]$, $|V_1(t, u_1) - V_1(t, u_2) \leq L_1 d[u_1, u_2]$, $L_1 > 0$, $V_1(t, u) \leq a_0(t, d[u, \hat{0}])$, *where* $a \in C[R_+ \times [0, p), R_+]$ *and* $a_0(t, \cdot) \in \mathcal{K}$ *for each* $t \in R_+$;

(A2) $D^+ V_1(t, u) \leq g_1(t, V_1(t, u))$, $(t, u) \in R^+ \times S(\rho)$, *where* $g_1 \in C[R_+^2, R]$ *and* $g_1(t, 0) \equiv 0$;

(A3) *for every* $\eta > 0$, *there exists a* $V_\eta \in C[R_+ \times S(\rho) \cap S^c(\eta), R_+]$,

$$|V_\eta(t, u_1) - V_\eta(t, u_2)| \leq L_\eta d[u_1, u_2],$$

$$b(d[u, \hat{0}]) \leq V(t, u) \leq a(d[u, \hat{0}]) a, b \in \mathcal{K},$$

and

$$D^+ V_1(t, u) + D^+ V_\eta(t, u) \leq g_2(t, V_1(t, u) + V_\eta(t, u))$$

for $(t, u) \in R_+ \times S(\rho) \cap S^c(\eta)$;

(A4) *the trivial solution* $w_1 \equiv 0$ *of*

$$w_1' = g_1(t, w_1), w_1(t_0) = w_{10} \geq 0, \tag{4.4.1}$$

is equi-stable;

(A5) *the trivial solution* $w_2 = 0$ *of*

$$w_2' = g_2(t, w_2), \quad w_2(t_0) = w_{20} \geq 0, \tag{4.4.2}$$

is uniformly stable.

Then the trivial solution of (4.2.1) is equi-stable.

Proof. Let $0 < \epsilon < \rho$ and $t_0 \in R_+$ be given. Since the trivial solution of (4.4.2) is uniformly stable, given $b(\epsilon) > 0$ and $t_0 \in R_+$, there exists a $\delta^0 = \delta^0(\epsilon) > 0$ satisfying

$$0 \leq w_{20} < \delta^0 \quad \text{implies} \quad w_2(t, t_0, w_{20}) < b(\epsilon), \quad t \geq t_0, \tag{4.4.3}$$

where $w_2(t, t_0, w_{20})$ is any solution of (4.4.2). In view of the hypothesis on $a(w)$, there is a $\delta_2 = \delta_2(\epsilon) > 0$ such that

$$a(\delta_2) < \frac{\delta^0}{2}. \tag{4.4.4}$$

Since the trivial solution of (4.4.1) is equi-stable, given $\frac{\delta^0}{2} > 0$ and $t_0 \in R_+$, we can find a $\delta^* = \delta^*(t_0, \epsilon) > 0$ such that

$$0 \leq w_{10} < \delta^* \quad \text{implies} \quad w_1(t, t_0, w_{10}) < \frac{\delta^0}{2}, \quad t \geq t_0, \tag{4.4.5}$$

where $w_1(t, t_0, w_{10})$ is any solution of (4.4.1).

Choose $w_{10} = V(t_0, u_0)$. Since $V_1(t, u) \leq a_0(t, d[u, \hat{0}])$, we see that there exists a $\delta_1 = \delta_1(t_0, \epsilon) > 0$ satisfying

$$d[u_0, \hat{0}] < \delta_1 \quad \text{and} \quad a_0(t_0, d[u_0, \hat{0}]) < \delta^*, \tag{4.4.6}$$

simultaneously. Define $\delta = \min(\delta_1, \delta_2)$. Then we claim that

$$d[u_0, \hat{0}] < \delta \quad \text{implies} \quad d[u(t), \hat{0}] < \epsilon, \quad t \geq t_0, \tag{4.4.7}$$

for any solution $u(t)$ of (4.2.1). If this is false, there would exist a solution $u(t)$ of (4.2.1) with $d[u_0, \hat{0}] < \delta$ and $t_1, t_2 > t_0$ such that

$$d[u(t_1), \hat{0}] = \delta_2, \quad d[u(t_2), \hat{0}] = \epsilon \quad \text{and} \quad \delta_2 \leq d[u(t), \hat{0}] \leq \epsilon \leq \rho \tag{4.4.8}$$

for $t_1 \leq t \leq t_2$. We let $\eta = \delta_2$ so that the existence of a V_η satisfying hypothesis (A3) is assured. Hence, setting

$$m(t) = V_1(t, u(t)) + V_\eta((t, u(t)), \quad t \in [t_1, t_2],$$

we obtain the differential inequality

$$D^+ m(t) \leq g_2(t, m(t)), \quad t_1 \leq t \leq t_2,$$

which yields

$$V_1(t_2, u(t_2)) + V_\eta(t_2, u(t_2)) \leq r_2(t_2, t_1, w_{20}), \tag{4.4.9}$$

where $w_{20} = V_1(t_1, u(t_1)) + V_\eta(t_1, u(t_1))$, $r_2(t, t_1, w_{20})$ is the maximal solution of (4.4.2). We also have, because of (A1) and (A2),

$$V_1(t_1, u(t_1)) \leq r_1(t_1, t_0, w_{10}),$$

with $w_{10} = V_1(t_0, u_0)$, where $r_1(t, t_0, w_{10})$ is the maximal solution of (4.4.1). By (4.4.5) and (4.4.6), we get

$$V_1(t_1, u(t_1)) < \frac{\delta_0}{2}. \qquad (4.4.10)$$

Also, by (4.4.4), (4.4.8) and (A3), we arrive at

$$V_\eta(t_1, u(t_1)) \le a(\delta_2) < \frac{\delta_0}{2}. \qquad (4.4.11)$$

Thus (4.4.10) and (4.4.11) and the definition of w_{20} shows that $w_{20} < \delta_0$ which, in view of (4.4.3), shows that $w_2(t_2, t_1, w_{20}) < b(\epsilon)$. It then follows from (4.4.9), $V_1(t, u) \ge 0$ and (A3),

$$b(\epsilon) = b(d[u(t_2), \hat{0}]) \le V_\eta(t_2, u(t_2)) \le r_2(t_2, t_1, w_{20}) < b(\epsilon).$$

This contradiction proves equi-stability of the trivial solution of (4.2.1) since (4.4.7) is then true.

The proof is complete.

The next result offers conditions for equi-asymptotic stability.

Theorem 4.4.2. *Let the assumptions of Theorem 4.4.1 hold except that condition (A2) is strengthened to*

(A2*) $D^+V_1(t, u) \le -c(w(t, u)) + g_1(t, V_1(t, (u)), c \in \mathcal{K}, w \in C[R_+ \times S(\rho), R_+],$
$|w(t, u_1) - w(t, u_2)| \le Nd[u_1, u_2], N > 0$ *and* $D^+w(t, u)$ *is bounded above or below.*

Then the trivial solution of (4.2.1) is equi-asymptotically stable, if $g_1(t, w)$ *is monotone nondecreasing in* w *and* $w(t, u) \ge b_0(d[u, \hat{0}]), b_0 \in \mathcal{K}.$

Proof.   By Theorem 4.4.1, the trivial solution of (4.2.1) is equi-stable. Hence letting $\epsilon = \rho$ so that $\delta_0 = \delta(\rho, t_0)$, we get, by equi-stability

$$d[u_0, \hat{0}] < \delta_0 \quad \text{implies} \quad d[u(t), \hat{0}] < \rho, \quad t \ge t_0.$$

We shall show that, for any solution $u(t)$ of (4.2.1) with $d[u_0, \hat{0}] < \delta_0$, it follows that $\lim_{t \to \infty} w(t, u(t)) = 0$, which implies by the property of $w(t, u)$, $\lim_{t \to \infty} d[u(t), \hat{0}] = 0$ and we are done.

Suppose that $\lim_{t \to \infty} \sup w(t, u(t)) \ne 0$. Then there would exist two divergent sequences $\{t_i'\}, \{t_i''\}$ and a $\sigma > 0$ satisfying

(a) $w(t_i', u(t_i')) = \frac{\sigma}{2}$, $w(t_i'', u(t_i'')) = \sigma$ and $w(t, u(t)) \ge \frac{\sigma}{2}$, $t \in (t_i'', t_i')$, or

(b) $w(t_i', u(t_i')) = \sigma$, $w(t_i'', u(t_i'')) = \frac{\sigma}{2}$ and $w(t, u(t)) \geq \frac{\sigma}{2}$, $t \in (t_i'', t_i')$.

Suppose that $D^+ w(t, u(t)) \leq M$. Then using (a) we obtain

$$\frac{\sigma}{2} = \sigma - \frac{\sigma}{2} = w(t_i'', u(t_i'')) - w(t_i', u(t_i')) \leq M(t_i'' - t_i'),$$

which shows that $t_i'' - t_i' \geq \frac{\sigma}{2M}$ for each i. Hence by (A$_2^*$) and Corollary 4.2.2, we have

$$V_1(t, u(t)) \leq r_1(t, t_0, w_{10}) - \sum_{i=1}^{n} \int_{t_i'}^{t_i''} C[w(s, u(s))]ds, \quad t \geq t_0.$$

Since $w_{10} = V_1(t_0, u_0) \leq a_0(t_0, d[u_0, \hat{0}]) \leq a_0(t_0, \delta_0) < \delta^*(\rho)$, we get from (4.4.5) $w_1(t, t_0, w_{10}) < \frac{\delta^0(\rho)}{2}$, $t \geq t_0$. We thus obtain

$$0 \leq V_1(t, u(t)) \leq \frac{\delta^0(\rho)}{2} - C\left(\frac{\sigma}{2}\right) \frac{\sigma}{2M} n.$$

For sufficiently larger n, we get a contradiction and therefore $\limsup_{t \to \infty} w(t, u(t)) = 0$. Since $w(t, u) \geq b_0(d[u, \hat{0}])$ by assumption, it follows that $\lim_{t \to \infty} d[u(t), \hat{0}] = 0$ and the proof is complete.

The following remarks are in order.

Remark 4.4.1. The functions $g_1(t, w) = g_2(t, w) \equiv 0$ are admissible in Theorem 4.4.1 so the same conclusion can be reached. If $V_1(t, u) \equiv 0$ and $g_1(t, w) \equiv 0$, then we get uniform stability from Theorem 4.4.1. If, on the other hand, $V_\eta(t, u) \equiv 0$, $g_2(t, w) \equiv 0$ and $V_1(t, u) \geq b(d[u, \hat{0}])$, $b \in \mathcal{K}$, then Theorem 4.4.1 yields equi-stability. We note that known results on equi-stability require the assumption to hold everywhere in $S(\rho)$ and Theorem 4.4.1 relaxes such a requirement considerably by the method of perturbing Lyapunov functions.

Remark 4.4.2. The functions $g_1(t, w) \equiv g_2(t, w) \equiv 0$ are admissible in Theorem 4.4.2 to yield equi-asymptotic stability. Similarly, if $V_\eta(t, u) \equiv 0$, $g_2(t, w) \equiv 0$ with $V_1(t, u) \geq b(d[u, \hat{0}])$, $b \in \mathcal{K}$, implies the same conclusion. If $V_1(t, u) \equiv 0$ and $g_1(t, w) \equiv 0$ in Theorem 4.4.1, to get uniform asymptotic stability, one needs to strengthen the estimate on $D^+ V_\eta(t, u)$. This we state as a corollary.

Corollary 4.4.1. *Assume that the assumptions of Theorem 4.4.1 hold with* $V_1(t,u) \equiv 0$, $g_1(t,w) \equiv 0$. *Suppose further that*

$$D^+V_\eta(t,u) \le -C[w(t,u)] + g_2(t, V_\eta(t,u)), (t,u) \in R_+ \times S(\rho) \cap S^c(\eta),$$
$$(4.4.12)$$

where $w \in C[R_+ \times S(\rho), R_+]$, $w(t,u) \ge b(d[u,\hat{0}])$, $c, b \in \mathcal{K}$ *and* $g_2(t,w)$ *is nondecreasing in* w. *Then the trivial solution of* (4.2.1) *is uniformly asymptotically stable.*

Proof.   The trivial solution of (4.2.1) is uniformly stable by Remark 4.4.1 in the present case. Hence taking $\epsilon = \rho$ and designating $\delta_0 = \delta(\rho)$, we have

$$d[u_0, \hat{0}] < \delta_0 \quad \text{implies} \quad d[u(t), \hat{0}] < \rho, \quad t \ge t_0.$$

To prove uniform attractivity, let $0 < \epsilon < \rho$ be given. Let $\delta = \delta(\epsilon) > 0$ be the number relative to ϵ in uniform stability. Choose $T = \frac{b(\rho)}{C(\delta)} + 1$. Then we shall show that there exists a $t^* \in [t_0, t_0 + T]$ such that $w(t^*, u(t^*)) < b(\delta)$ for any solution $u(t)$ of (4.2.1) with $d[u_0, \hat{0}] < \delta_0$. If this is not true, $w(t, u(t)) \ge b(\delta)$, $t \in [t_0, t_0 + T]$. Now using the assumption (4.4.12) and arguing as in Corollary 4.2.2, we get

$$0 \le V_\eta(t_0 + T, u(t_0 + T)) \le r_2(t_0 + T, t_0, w_{20}) - \int_{t_0}^{t_0+T} w(s, u(s))ds.$$

This yields, since $r_2(t, t_0, w_{20}) < b(\rho)$ and the choice of T,

$$0 \le V_\eta(t_0 + T, u(t_0 + T)) \le b(\rho) - c(\delta)T < 0,$$

which is a contradiction. Hence there exists a $t^* \in [t_0, t_0 + T]$ satisfying $w(t^*, u(t^*)) < b(\delta)$, which implies $d[u(t^*), \hat{0}] < \delta$. Consequently, it follows, by uniform stability that

$$d[u_0, \hat{0}] < \delta_0 \quad \text{implies} \quad d[u(t), \hat{0}] < \epsilon, \quad t \ge t_0 + T,$$

and the proof is complete.

4.5   Criteria for Boundedness

We shall, in this section, investigate the boundedness of solutions of the fuzzy differential equation

$$u' = f(t,u), \quad u(t_0) = u_0, \tag{4.5.1}$$

where $f \in C[R_+ \times E^n, E^n]$. Corresponding to the definitions of various stability notions given in Section 3.9, we also have boundedness concepts, which we shall define.

Definition 4.5.1. *The solutions of* (4.5.1) *are said to be*

(B1) *equi-bounded, if for any $\alpha > 0$ and $t_0 \in R_+$, there exists a $\beta = \beta(t_0, \alpha) > 0$ such that*

$$d[u_0, \hat{0}] < \alpha \quad implies \quad d[u(t), \hat{0}] < \beta, \quad t \geq t_0;$$

(B2) *uniform-bounded, if β in* (B1) *does not depend on t_0;*

(B3) *quasi-equi-ultimately bounded for a bound B, if there exists a $B > 0$ and a $T = T(t_0, \alpha) > 0$ such that*

$$d[u_0, \hat{0}] < \alpha \quad implies \quad d[u(t), \hat{0}] < B, \quad t \geq t_0 + T;$$

(B4) *quasi-uniform ultimately bounded, if T in* (B3) *is independent of t_0;*

(B5) *equi-ultimately bounded, if* (B1) *and* (B3) *hold simultaneously;*

(B6) *uniform-ultimately bounded if* (B2) *and* (B4) *hold simultaneously;*

(B7) *equi-Lagrange stable, if* (B1) *and* (S3) *hold;*

(B8) *uniformly Lagrange stable, if* (B2) *and* (S4) *hold.*

Using the comparison results of Section 4.2, we shall prove simple boundedness results.

Theorem 4.5.1. *Assume that*

(i) $V \in C[R_+ \times E^n, R_+]$, $|V(t, u_1) - V(t, u_2)| \leq L d[u_1, u_2]$, $L > 0$ *and for* $(t, u) \in R_+ \times E^n$, $D^+ V(t, u) \leq 0$;

(ii) $b(d[u, \hat{0}]) \leq V(t, u) \leq a(t, d[u_0, \hat{0}])$ *for* $(t, u) \in R_+ \times E^n$, $a(t, \cdot), b \in \mathcal{K}$, *where* $\mathcal{K} = [\sigma \in C[R_+, R_+] : \sigma(w)$ *is increasing and* $\sigma(w) \to \infty$ *as* $w \to \infty]$.

Then (B1) *holds.*

Proof.   Let $0 < \alpha$ and $t_0 \in R_+$ be given. Choose $\beta = \beta(t_0, \alpha)$ such that

$$a(t_0, \alpha) < b(\beta). \tag{4.5.2}$$

With this β, (B_1) holds. If this is not ture, there would exist a solution $u(t) = u(t, t_0, u_0)$ of (4.5.1) and a $t_1 > t_0$ such that

$$d[u(t_1), \hat{0}] = \beta \quad \text{and} \quad d[u(t), \hat{0}] \le \beta, \quad t_0 \le t \le t_1.$$

Assumption (i) and Corollary 4.2.1 show that

$$V(t, u(t)) \le V(t_0, u_0), \quad t_0 \le t \le t_1.$$

As a result, condition (ii) and (4.5.2) yield

$$
\begin{aligned}
b(\beta) &= b(d[u(t_1), \hat{0}]) \le V(t_1, u(t_1)) \le V(t_0, u_0) \\
&\le a(t_0, d[u_0, \hat{0}]) < a(t_0, \alpha) < b(\beta).
\end{aligned}
$$

This contradiction proves (B1) and we are done.

For uniform boundedness, we have the following result, whose assumptions are weaker.

Theorem 4.5.2. *Assume that*

(i) $V \in C[R_+ \times S^c(\rho, R_+]$, *where ρ may be large,* $|V(t, u_1) - V(t, u_2)| \le Ld[u_1, u_2]$, *and* $D^+V(t, u) \le 0$ *for* $R_+ \times S^c(\rho)$;

(ii) $b(d[u, \hat{0}]) \le V(t, u) \le a(d[u, \hat{0}])$, $a, b \in \mathcal{K}$, *whose a, b are defined only on* $[\rho, \infty)$.

Then (B2) *is true.*

Proof.   The proof is similar to the proof of Theorem 4.5.1 except that the choice of β is now made so that $a(\alpha) < b(\beta)$ and consequently β is independent of t_0. Also, $\alpha > \rho$ for the proof since the assumptions are only for $S^c(\rho)$. However, if $0 < \alpha \le \rho$, we can take $\beta = \beta(\rho)$ and hence there is no problem.

We shall give a typical result that offers conditions for equi-ultimate boundedness, that is, (B5).

Theorem 4.5.3. *Let all the assumptions of Theorem 4.5.1 hold except that we strengthen the estimate on $D^+V(t, u)$ as*

$$D^+V(t, u) \le -\eta V(t, u), \quad \eta > 0, \quad (t, u) \in R_+ \times E^n, \tag{4.5.3}$$

and suppose that condition (ii) *holds for* $d[u, \hat{0}] \ge B$. *Then* (B5) *holds.*

Proof.   Clearly (B1) is satisfied by Theorem 4.5.1. Hence

$$d[u_0, \hat{0}] < \alpha \quad \text{implies} \quad d[u(t), \hat{0}] < \beta, \quad t \geq t_0.$$

Now (4.5.3) yields the estimate

$$V(t, u(t)) \leq V(t_0, u_0)e^{-\eta(t-t_0)} \quad t \geq t_0. \tag{4.5.4}$$

Let $T = \frac{1}{\eta} \ln \frac{a(t_0, \alpha)}{b(B)}$ and suppose that for $t \geq t_0 + T$, $d[u(t), \hat{0}] \geq B$. Then we get from (4.5.4)

$$b(B) \leq b(d[u(t), \hat{0}]) \leq V(t, u(t)) \leq a(t_0, \alpha)e^{-\eta T} < b(B).$$

This contradiction proves (B5) and the proof is complete.

Finally we shall offer a result providing nonuniform boundedness property utilizing the method of perturbing Lyapunov functions.

Theorem 4.5.4. *Assume that*

(i) $\rho > 0, V_1 \in C[R_+ \times S(\rho), R_+], V_1$ *is bounded for* $(t, u) \in R_+ \times \partial S(\rho)$, *and*

$$|V_1(t, u_1) - V(t, u_2)| \leq L_1 d[u_1, u_2], \quad L_1 > 0,$$

$$\begin{aligned}
D_+V_1(t, u) &= \limsup_{h \to 0^+} \frac{1}{h}[V_1(t + h, u + hf(t, u)) - V_1(t, u)] \\
&\leq g_1(t, V_1), \quad (t, u) \in R^+ \times S^c(\rho)
\end{aligned}$$

where $g_1 \in C[R_+^2, R]$;

(ii) $V_2 \in C[R_+ \times S^c(\rho), R_+]$,

$$b(d[u, \hat{0}]) \leq V_2(t, u) \leq a(d[u, \hat{0}]), \quad a(\cdot), b(\cdot) \in K$$

$$D^+V_1 + D^+V_2 \leq g_2(t, V_1(t, u) + V_2(t, u)) \quad g_2 \in C[R_+^2, R]$$

(iii) *the scalar differential equation*

$$w_1' = g_1(t, w_1), \quad w_1(t_0) = w_{10} \geq 0 \tag{4.5.5}$$

and

$$w_2' = g_2(t, w_2) \quad w_2(t_0) = w_{20} \geq 0 \tag{4.5.6}$$

are equi-bounded and uniformly bounded respectively.

Then the system (4.5.1) is equi-bounded.

Proof.   Let $B_1 > \rho$ and $t_0 \in R_+$ be given. Let $\alpha_1 = \alpha_1(t_0, B_1) = \max(\alpha_0, \alpha^*)$, where $\alpha_0 = \max[V_1(t_0, u_0) : u_0 \in cl\{S(B_1) \cap S^c(\rho)\}]$ and $\alpha^* \geq V_1(t, u)$ for $(t, u) \in R^+ \times \partial S(\rho)$. Since equation (4.5.5) is equi-bounded, given $\alpha_1 > 0$, and $t_0 \in R_+$, there exists a $\beta_0 = \beta_0(t_0, \alpha_1)$, such that

$$w_1(t, t_0, w_{10}) < \beta_0 \quad t \geq t_0 \qquad\qquad (4.5.7)$$

provided $w_{10} < \alpha_1$, where $w_1(t, t_0, w_{10})$ is any solution of (10). Let $\alpha_2 = a(B_1) + \beta_0$, then uniform boundedness of equation (4.5.6) yields that

$$w_2(t, t_0, w_{20}) < \beta_1(\alpha_2), \quad t \geq t_0, \qquad\qquad (4.5.8)$$

provided $w_{20} < \alpha_2$, where $w_2(t, t_0, w_{20})$ is any solution of (4.5.6). Choose B_2 satisfying

$$b(B_2) > \beta_1(\alpha_2). \qquad\qquad (4.5.9)$$

We now claim that $u_0 \in S(B_1)$ implies that $u(t, t_0, u_0) \in S(B_2)$ for $t \geq t_0$, where $u(t, t_0, u_0)$ is any solution of (4.5.1).

If it is not true, there exists a solution $u(t, t_0, u_0)$ of (4.5.7) with $u_0 \in S(B_1)$, such that for some $t^* > t_0, d[u(t^*, t_0, u_0), \hat{0}] = B_2$. Since $B_1 > \rho$, there are two possibilities to consider:

(1) $u(t, t_0, u_0) \in S^c(\rho)$ for $t \in [t_0, t^*]$;

(2) there exists a $\bar{t} \geq t_0$ such that $u(\bar{t}, t_0, u_0) \in \partial S(\rho)$ and $u(t, t_0, u_0) \in S^c(\rho)$ for $t \in [\bar{t}, t^*]$.

If (1) holds, we can find $t_1 > t_0$, such that

$$\begin{aligned} u(t_1, t_0, u_0) &\in \partial S(B_1), \\ u(t^*, t_0, u_0) &\in \partial S(B_2), \quad \text{and} \\ u(t, t_0, u_0) &\in S^c(B_1), \quad t \in [t_1, t^*]. \end{aligned} \qquad (4.5.10)$$

Setting $m(t) = V_1(t, u(t, t_0, u_0)) + V_2(t, u(t, t_0, u_0))$ for $t \in [t_1, t^*]$, then using Theorem 4.2.1, we can obtain the differential inequality

$$D^+ m(t0 \leq g_2(t, m(t)) \quad t \in [t_1, t^*],$$

and so

$$m(t) \leq \gamma_2(t, t_1, m(t_1)) \quad t \in [t_1, t^*]$$

where $\gamma_2(t, t_1, v_0)$ is the maximal solution of (4.5.6) with $\gamma_2(t_1, t_1, v_0) = v_0$. Thus

$$V_1(t^*, u(t^*, t_0, u_0)) + V_2(t^*, u(t^*, t_0, u_0)) \qquad (4.5.11)$$

$$\leq \gamma_2(t^*, t_1, V_1(t_1, u(t_1, t_0, u_0)) + V_2(t_1, u(t_1, t_0, u_0))).$$

Similarly, we also have

$$V_1(t_1, u(t_1, t_0, u_0)) \leq \gamma_1(t_1, t_0, V_1(t_0, u_0)) \qquad (4.5.12)$$

where $\gamma_1(t, t_0, u_0)$ is the maximal solution of (4.5.5). Set $w_{10} = V_1(t_0, u_0) < \alpha_1$. Then

$$V_1(t_1, u(t_1, t_0, u_0)) \leq \gamma_1(t_1, t_0, V_1(t_0, u_0)) \leq \beta_0$$

since (4.5.7) holds. Furthermore, $V_2(t_1, u(t_1, t_0, u_0)) \leq a(B_1)$ and (4.5.10). Consequently, we have

$$\begin{aligned} w_{20} &= V_1(t_1, u(t_1, t_0, u_0)) + V_2(t_1, u(t_1, u(t_1, t_0, u_0)) \\ &\leq \beta_0 + a(B_1) = \alpha_2. \end{aligned} \qquad (4.5.13)$$

Combining (4.5.8), (4.5.9), (4.5.10) and (4.5.13), we obtain

$$b(B_2) \leq m(t^*) \leq \gamma(t^*) \leq \beta_1(\alpha_2) < b(B_2), \qquad (4.5.14)$$

which is a contradiction.

If case (2) holds, we also arrive at the inequality (4.5.11), where $t_1 > \bar{t}$ satisfies (4.5.10). We now have, in place of (4.5.12), the relation

$$V_1(t_1, u(t_1, t_0, u_0)) \leq \gamma_1(t_1, \bar{t}, V_1(\bar{t}, u(\bar{t}, t_0, x_0))).$$

Since $u(\bar{t}, t_0, u_0) \in \partial S(\rho)$ and $V_1(\bar{t}, u(\bar{t}, t_0, x_0)) \leq \alpha^* \leq \alpha_1$, arguing as before, we get the contradiction (4.5.14). This proves that

for any given $B_1 > \rho, t_0 > 0$, there exists a B2 such that $u_0 \in S(B1)$ implies $u(t, t_0, u_0) \in S(B_2), t \geq t_0$.

For $B_1 < \rho$, we set $B_2(t_0, B_1) = B_2(t_0, \rho)$ and hence the proof is complete.

4.6 Fuzzy Differential Systems

Recall that we have so far been discussing the fuzzy differential equation

$$u' = f(t, u), \quad u(t_0) = u_0,$$

---

where $f \in C[R_+ \times E^n, E^n]$, which corresponds, without fuzziness, to scalar differential equations. To consider the situation analogous to differential systems, we need to prepare appropriate notation. In this section, we shall therefore attempt to consider the fuzzy differential system, given by

$$U' = F(t,U), \quad U(t_0) = U_0, \tag{4.6.1}$$

where $F \in C[R_+ \times E^{nN}, E^{nN}], U \in E^{nN}, E^{nN} = (E^n \times E^n \times \cdots \times E^n, N$ times), $U = (u_1, u_2, ..., u_N)$ such that for each $i, 1 \le i \le N$, $u_i \in E^n$. Note also $U_0 \in E^{nN}$.

We have two possibilities to measure the new variables U, U_0, F, that is,

(1) we can define $d_0[U,V] = \sum_{i=1}^{N} d[u_i, v_i]$, where $U, V \in E^{nN}$ and employ the metric space (E^{nN}, d_0), or

(2) define the generalized metric space (E^{nN}, D), where $D \in R_+^N$ such that $D[U,V] = (d[u_1,v_1], d[u_2,v_2], ..., d[u_N,v_N])$.

In this framework, the assumption (3.4.1) of Theorem 3.4.1 appears as

$$d_0[F(t,U), F(t,V)] = \sum_{i=1}^{N} d[f_i(t,U), f_i(t,V)] \le g(t, d_0[U,V]) \tag{4.6.2}$$

if we utilize option (1) above. On the other hand, if we choose option (2), assumption (3.4.1) will be of the form

$$D[F(t,U), F(t,V)] \le G(t, D[U,V]), \tag{4.6.3}$$

where $G \in C[R_+ \times R_+^N, R^N]$. In this case, condition (3.2.3) reduces to

$$D[F(t,U), F(t,V)] \le SD[U,V] \tag{4.6.4}$$

where $S = (S_{ij})$ is an $N \times N$ matrix with $S_{ij} \ge 0$, for all i, j, which corresponds to the generalized contractive condition. Of course, the matrix S needs to satisfy a suitable condition, that is, for some $k > 1$, S^k must be an A-matrix, which means $I - S^k$ is positive definite, where I is the identity matrix. For details of generalized spaces and contraction mapping theorem in this setup see Bernfeld and Lakshmikantham [4], page 226.

Moreover, in order to arrive at the corresponding estimate (3.4.3) of Theorem 3.4.1, for example, one is required to utilize the corresponding theory of systems of differential inequalities, which demands that $G(t,w)$ have the quasi-monotone property, which is defined as follows:

$$w_1 \le w_1 \quad \text{and} \quad w_{1i} = w_{2i} \quad \text{for some} \quad i, 1 \le i \le N, \quad \text{implies}$$

$$G_i(t, w_1) \leq G_i(t, w_2), \quad w_1, w_2 \in R^N.$$

If $G(t, w) = Aw$, where A is an $N \times N$ matrix, then the quasi-monotone property reduces to requiring $a_{ij} \geq 0, i \neq j$.

The method of vector Lyapunov-like functions has been very effective in the investigation of the qualitative properties of large-scale differential systems. We shall extend this technique to fuzzy differential systems (4.6.1), where, as we shall see, both metrics described above are very useful. For this purpose, let us prove the following comparison result in terms of vector Lyapunov-like functions relative to the fuzzy differential system (4.6.1). We note that the inequalities between vectors in R^N are to be understood as componentwise.

Theorem 4.6.1. *Assume that* $V \in C[R_+ \times E^{nN}, R_+^N]$, $|V(t, U_1) - V(t, U_2)| \leq AD[U_1, U_2]$, *where* A *is an* $N \times N$ *matrix with nonnegative elements, and for* $(t, U) \in R_+ \times E^{nN}$

$$D^+ V(t, U) \leq G(t, V(t, U)) \tag{4.6.5}$$

where $G \in C[R_+ \times R_+^N, R^N]$. *Suppose further that* $G(t, w)$ *is quasi-monotone in* w *for each* $t \in R_+$ *and* $r(t) = r(t, t_0, w_0)$ *is the maximal solution of*

$$w' = G(t, w), \quad w(t_0) = w_0 \geq 0 \tag{4.6.6}$$

existing for $t \geq t_0$. *Then*

$$V(t, U(t)) \leq r(t), \quad t \geq t_0, \tag{4.6.7}$$

where $U(t)$ *is any solution of* (4.6.1) *existing for* $t \geq t_0$.

Proof. Let $U(t)$ be any solution of (4.6.1) existing for $t \geq t_0$. Define $m(t) = V(t, U(t))$ so that $m(t_0) = V(t_0, U_0) \leq w_0$. Now for small $h > 0$, we have, in view of Lipschitz conditions,

$$
\begin{aligned}
m(t + h) - m(t) &= V(t + h, U(t + h)) - V(t, U(t)) \\
&\leq AD[U(t + h), U(t) + hF(t, U(t))] \\
&\quad + V(t + h, U(t) + hF(t, U(t))) - V(t, U(t)).
\end{aligned}
$$

It follows therefore that

$$
\begin{aligned}
D^+ m(t) &= \limsup_{h \to 0^+} \frac{1}{h}[m(t + h) - m(t)] \leq D^+ V(t, U(t)) \\
&\quad + A \limsup_{h \to 0^+} \frac{1}{h}[D[U(t + h), U(t) + hF(t, U(t))]].
\end{aligned}
$$

Since $U'(t)$is assumed to exist, we see that $U(t+h) = U(t) + Z(t)$ where $Z(t)$ is the H-difference for small $h > 0$. Hence utilizing the properties of $D[U,V]$, we obtain

$$
\begin{aligned}
D[U(t+h), U(t) + hF(t, U(t))] &= D[U(t) + Z(t) \\
&< U(t) + hF(t, U(t))] \\
&= D[Z(t), hF(t, U(t))] \\
&= D[U(t+h) - U(t), hF(t, U(t))].
\end{aligned}
$$

As a result, we get

$$
\frac{1}{h} D[U(t+h), U(t) + hF(t, U(t))] = D\left[\frac{U(t+h) - U(t)}{h}, F(t, U(t)) \right]
$$

and consequently

$$
\lim_{h \to 0+} \sup \frac{1}{h} [D[U(t+h) - U(t) + hF(t, U(t))]
$$

$$
\begin{aligned}
&= \lim_{h \to 0+} \sup \frac{1}{h} \left[D\left[\frac{U(t+h) - U(t)}{h}, F(t, U(t)) \right] \right] \\
&= D[U'(t), F(t, U(t))] \\
&= 0,
\end{aligned}
$$

since $U(t)$ is a solution of (4.6.1). We therefore have the vectorial differential inequality

$$
D^+ m(t) \leq G(t, m(t)), \quad m(t_0) \leq w_0, \quad t \geq t_0,
$$

which by the theory of differential inequalities for systems (Lakshmikantham and Leela [61]) yields

$$
m(t) \leq r(t), \quad r \geq t_0,
$$

proving the claimed estimate (4.6.7).

The following corollary of Theorem 4.6.1 is interesting.

Corollary 4.6.1. *The function $G(t, w) = Aw$, where A is an $N \times N$ matrix satisfying $a_{ij} \geq 0, i \neq j$, is admissible in Theorem 4.6.1 and yields the estimate*

$$
V(t, U(t)) \leq V(t_0, U_0) e^{A(t-t_0)}, \quad t \geq t_0.
$$

4.7 The Method of Vector Lyapunov Functions

We shall prove a typical result that gives sufficient conditions in terms of vector Lyapunov-like functions for the stability properties of the trivial solution of the fuzzy differential system (4.6.1).

Theorem 4.7.1. *Assume that*

(i) $G \in C[R_+ \times R_+^N, R^N]$, $G(t,0) \equiv 0$ *and* $G(t,w)$ *is quasi-monotone nondecreasing in* w *for each* $t \in R_+$;

(ii) $V \in C[R_+ \times S(\rho), R_+^N]$, $|V(t,U_1) - V(t,U_2)| \leq AD[U_1, U_2]$, *where* A *is a nonnegative* $N \times N$ *matrix and the function*

$$V_0(t,U) = \sum_{i=1}^{N} V_i(t,U) \qquad (4.7.1)$$

satisfies

$$b(d_0[U, \hat{0}]) \leq V_0(t,U) \leq a(d_0[U, \hat{0}]), \quad a, b \in \mathcal{K};$$

(iii) $F \in C[R_+ \times S(\rho), E^{nN}]$, $F(t, \hat{0}) \equiv \hat{0}$ *and*

$$D^+V(t,U) \leq G(t, V(t,U)), \quad (t,U) \in R_+ \times S(\rho),$$

where $S(\rho) = [U \in E^{nN} : d_0[U, \hat{0}] < \rho]$.

Then, the stability properties of the trivial solution of (4.6.6) *imply the corresponding stability properties of the trivial solution of* (4.6.1).

Proof. We shall prove only equi-asymptotic stability of the trivial solution of (4.6.1). For this purpose, let us first prove equi-stability. Let $0 < \epsilon < \rho$ and $t_0 \in R_+$ be given. Assume that the trivial solution of (4.6.6) is equi-asymptotically stable. Then it is equi-stable. Hence given $b(\epsilon) > 0$ and $t_0 \in R_+$, there exists a $\delta_1 = \delta_1(t_0, \epsilon) > 0$ such that

$$\sum_{i=1}^{N} w_{i0} < \delta_1 \quad \text{implies} \quad \sum_{i=1}^{N} w_i(t, t_0, w_0) < b(\epsilon), \quad t \geq t_0, \qquad (4.7.2)$$

where $w(t, t_0, w_0)$ is any solution of (4.6.6). Choose $w_0 = V(t_0, U_0)$ and a $\delta = \delta(t_0, \epsilon) > 0$ satisfying

$$a(\delta) < b(\epsilon). \qquad (4.7.3)$$

Let $d_0[U_0, \hat{0}] < \delta$. Then we claim that $d_0[U(t), \hat{0}] < \epsilon, t \geq t_0$, for any solution $U(t) = U(t, t_0, U_0)$ of (4.6.1). If this is not true, there would exist a solution $U(t)$ of (4.6.1) with $d_0[U_0, \hat{0}] < \delta$ and a $t_1 > t_0$ such that

$$d_0[U(t_1), \hat{0}] = \epsilon \quad \text{and} \quad d_0[U(t), \hat{0}] \leq \epsilon < \rho, \quad t_0 \leq t \leq t_1. \tag{4.7.4}$$

Hence we have by Theorem 4.6.1

$$V(t, U(t)) \leq r(t, t_0, w_0), \quad t_0 \leq t \leq t_1, \tag{4.7.5}$$

where $r(t, t_0, w_0)$ is the maximal solution of (4.6.6). Since

$$V_0(t_0, U_0) \leq a(d_0[U_0, \hat{0}]) < a(\delta) < \delta_1,$$

the relations (4.7.2), (4.7.3), (4.7.4) and (4.7.5) yield

$$b(\epsilon) \leq V_0(t_1, U(t_1)) \leq r_0(t_1, t_0, w_0) < b(\epsilon),$$

where $r_0(t, t_0, w_0) = \sum_{i=1}^{N} r_i(t, t_0, w_0)$. This contradiction proves that the trivial solution of (4.6.1) is equi-stable.

Suppose next that the trivial solution of (4.6.6) is quasi-equi-asymptotically stable. Set $\epsilon = \rho$ and $\hat{\delta}_0 = \delta(t_0, \rho)$. Let $0 < \eta < \rho$. Then given $b(\eta)$ and $t_0 \in R_+$, there exist $\delta_1^* = \delta_1(t_0) > 0$ and $T = T(t_0, \eta) > 0$ satisfying

$$\sum_{i=1}^{N} w_{i0} < \delta_0^* \quad \text{implies} \quad \sum_{i=1}^{N} w_i(t, t_0, w_0) < b(\eta), \quad t \geq t_0 + T. \tag{4.7.6}$$

Choosing $w_0 = V(t_0, U_0)$ as before, we find $\delta_0^* = \delta_0(t_0) > 0$ such that $a(\delta_0^*) < \delta_1^*$. Let $\delta_0 = \min(\delta_1^*, \delta_0^*)$ and $d_0[U_0, \hat{0}] < \delta_0$. This implies $d_0[U(t), \hat{0}] < \rho, t \geq t_0$ and therefore the estimate (4.7.5) holds for all $t \geq t_0$. Suppose now that there is a sequence $\{t_k\}, t_k \geq t_0 + T, t_k \to \infty$ as $k \to \infty$, and $\eta \leq d_0[U(t_k), \hat{0}]$, where $U(t)$ is any solution of (4.6.1) with $d_0[U_0, \hat{0}] < \delta_0$. In view of (4.7.6), this leads to the contradiction

$$b(\eta) \leq V_0(t_k, U(t_k)) \leq r_0(t_k, t_0, w_0) < b(\eta).$$

Hence the trivial solution of (4.6.1) is equi-asymptotically stable and the proof is complete.

In order to apply the method of vector Lyapunov functions to concrete problems, it is necessary to know the properties of the solutions of the comparison system (4.6.6), which is difficult in general, except when $G(t, w) = Aw$, where A is a quasi-monotone $N \times N$ stability matrix. Hence we shall present some simple and useful techniques to deal with this problem.

We shall first prove a result which reduces the study of the properties of solutions of (4.6.6) to that of a scalar differential equation

$$v' = G_0(t, v), \quad v(t_0) = v_0 \geq 0, \tag{4.7.7}$$

where $g_0 \in C[R_+^2, R]$. Specifically we have the following result.

Lemma 4.7.1. *Assume that* $L \in C^1[R_+, R_+^N]$, $G \in C[R_+ \times R_+^N, R^N]$, $G_0 \in C[R_+^2, R]$ *and* G, G_0 *are smooth enough to assure existence and uniqueness of solutions for* $t \geq t_0$ *of* (4.6.6) *and* (4.7.7) *respectively. Suppose further that for* $(t, v) \in R_+^2$,

$$G(t, L(v)) \leq \frac{dL(v)}{dv} G(t, v).$$

Then $w_0 \leq L(v_0)$ *implies*

$$w(t, t_0, w_0) \leq L(v(t, t_0, v_0)), \quad t \geq t_0, \tag{4.7.8}$$

where $w(t, t_0, w_0), v(t, t_0, v_0)$ *are the solutions of* (4.6.6) *and* (4.7.7) *respectively.*

Proof. Set $m(t) = L(v(t, t_0, v_0))$ so that $m(t_0) = L(v_0) \geq w_0$ and

$$
\begin{aligned}
m'(t) &= \frac{dL(v(t, t_0, v_0))}{dv} G_0(t, v(t, t_0, v_0)) \\
&\geq G(t, L(v(t, t_0, v_0))) = g(t, m(t)).
\end{aligned}
$$

Hence by the comparison Theorem 1.4.1 in Lakshmikantham and Leela [61], we get the stated result in view of uniqueness of solutions.

Let us give an example to illustrate Lemma 4.7.1. Suppose that $G_1 = -2w_1^2$, $G_2 = -2w_2^3 + 2w_1 w_2^{\frac{3}{2}}$ so that

$$w_1' = -2w_1^2,$$

$$w_2' = -2w_2^3 + 2w_1 w_2^{\frac{3}{2}}. \tag{4.7.9}$$

Choosing $L_1(v) = \frac{3}{5} v^{\frac{3}{2}}$, $L_2(v) = v$ and

$$G_0(t, v) = \begin{cases} -\frac{2}{5} v^3, & 0 \leq v < 1, \\ -\frac{2}{5} v^{\frac{5}{2}}, & 1 \leq v, \end{cases}$$

the assumptions of Lemma 4.7.1 are satisfied. Clearly the trivial solution of (4.7.7) is uniformly asymptotically stable and therefore the trivial solution of (4.7.9) is also uniformly asymptotically stable.

Lemma 4.7.2. *Assume that*

$$Q \in C^1[R_+^N, R_+], \quad G \in C[R_+ \times R_+^N, R^N], \quad G_0 \in C[R_+^2, R]$$

and for $(t, w) \in R_+ \times R_+^N$,

$$\frac{dQ(w)}{dw} G(t, w) \leq G_0(t, Q(w)). \tag{4.7.10}$$

Then any solution $w(t) = w(t, t_0, w_0)$ *of (4.6.6) existing for* $t \geq t_0$, *satisfies*

$$Q(w(t)) \leq v(t), \quad t \geq t_0$$

where $v(t) = v(t, t_0, v_0)$ *is the maximal solution (2.5.3) existing for* $t \geq t_0$, *provided* $Q(w_0) \leq v_0$.

Proof.   Let $w(t) = w(t, t_0, w_0)$ be any solution of (4.6.6) existing for $t \geq t_0$. Set $p(t) = Q(w(t))$. Then we have

$$p'(t) = \frac{dQ(w(t))}{dw} G(t, w(t)) \leq G_0(t, Q(w(t))) = G_0(t, p(t))),$$

and $p(t_0) \leq v_0$. Hence by Theorem 1.4.1 in Lakshmikantham and Leela [60], it follows that $p(t) \leq v(t), t \geq t_0$, where $v(t)$ is the maximal solution of (2.5.3). Hence the proof is complete.

As an example, consider the case $G(t, w) = Aw$ where A is an $N \times N$ matrix with $a_{ij} \geq 0, i \neq j$, and A is quasi-diagonally dominant, that is, for some $d_i > 0$,

$$d_i |a_{ii}| > \sum_{\substack{j=1 \\ i \neq j}} d_j |a_{ij}|. \tag{4.7.11}$$

Choosing $Q(w) = \sum_{i=1}^N d_i w_i$ for some $d_i > 0$, we see that (4.7.10) is satisfied by $G_0(t, v) = -\gamma v$, for some $\gamma > 0$ in view of (4.7.11). Consequently, the trivial solution of (4.7.7) is exponentially asymptotically stable which implies that the trivial solution of (4.6.6) does have the same property.

4.8   Linear Variation of Parameters Formula

Let us consider the linear fuzzy differential system

$$U' = AU, \quad U(t_0) = U_0, \quad t_0 \geq 0 \tag{4.8.1}$$

where A is an $N \times N$ matrix of reals and $U = (u_1, u_2, ..., u_N)$ such that for each $i, 1 \leq i \leq N, u_i \in E^n$. Note also $U_0 \in E^{nN}$. See Section 4.6 for notation. We shall also consider the following fuzzy differential system which is a perturbation of (4.8.6), namely,

$$U' = AU + F(t, U), \quad U(t_0) = v_0, \quad t_0 \geq 0 \qquad (4.8.2)$$

where $F \in C[R_+ \times E^{nN}, E^{nN}]$. We recall that

$$D[U, V] = (d[u_1, v_1], d[u_2, v_2], ..., d[u_N, v_N])$$

is the generalized metric and $D[U, V] \in R_+^N$. Also, the real numbers can be embedded in E^1 by the correspondence

$$a \to a(t) = \begin{cases} 1 & \text{if } t = a, \\ 0 & \text{elsewhere.} \end{cases} \qquad (4.8.3)$$

Then we can generalize multiplication by a real number and for any real number a, we get

$$[au]^\alpha = a[u]^\alpha, \quad 0 \leq \alpha \leq 1, \quad u \in E^n.$$

We also know that $[u+v]^\alpha = [u]^\alpha + [v]^\alpha, u, v \in E^n$. Thus we can write (4.8.1) in the expanded form

$$u_i' = a_{i1}u_1 + a_{i2}u_2 + \cdots + a_{iN}u_N \qquad (4.8.4)$$

for each $i, 1 \leq i \leq N$. As a result, we can estimate $D[AU, AV]$ where $U = (u_1, u_2, \cdots, u_N)$, $V = (v_1, v_2, \cdots, v_N)$ such that $u_i, v_i \in E^n$ as follows: For each i

$$d_H \left[\left[\sum_{j=1}^N a_{ij}u_j \right]^\alpha, \left[\sum_{j=1}^n a_{ij}v_j \right]^\alpha \right]$$

$$\begin{aligned} &= d_H \left[[a_{ij}u_1]^\alpha, \cdots, [a_{iN}u_N]^\alpha, [a_{i1}v_1]^\alpha, \cdots [a_{iN}v_N]^\alpha \right] \\ &= d_H \left[a_{i1}[u_1]^\alpha, \cdots, d_{iN}[u_N]^\alpha, a_{i1}[v_1]^\alpha, \cdots, a_{iN}[v_N]^\alpha \right] \\ &\leq |a_{i1}|d_H \left[[u_1]^\alpha, [v_1]^\alpha \right] + |a_{i2}|d_H \left[[u_2]^\alpha, [v_2]^\alpha + \cdots + |a_{iN}|d_H [u_N]^\alpha, [v_N]^\alpha \right]. \end{aligned}$$

Here we have used carefully the properties of d_H. Since this estimation is valid for each i, we obtain

$$D[AU, AV] \leq QD[U, V] \qquad (4.8.5)$$

where $Q = |a_{ij}|, i, j = 1, 2, ..., N$, is the $N \times N$ matrix with nonnegative elements $|a_{ij}|$. Recall that (4.8.5) is a vectorial inequality with componentwise inequalities. Assume that for some $k > 1$, $I - Q^k$ is positive definite where I is the identity matrix. Then the generalized contraction mapping theorem (see Bernfeld and Lakshmikantham [4], p. 226) assures the existence of the unique solution to the IVP (4.8.1). One can verify easily that $U(t) = e^{A(t-t_0)}v_0$ is the unique solution of (4.8.1). Consequently, the variation of parameters formula relative to the IVP (4.8.2) takes the usual form

$$U(t) = d^{A(t-t_0)}v_0 + \int_{t_0}^{t} e^{A(t-s)} F(s, U(s))ds, \quad t \geq t_0. \qquad (4.8.6)$$

We shall employ the variation of parameters formula (4.8.6) later in Chapter 5, in discussing the boundary value problem.

Let us now consider the fuzzy linear homogeneous system with periodic boundary condition

$$U'(t) = AU(t) + \sigma(t), \quad U(0) = U(2\pi), \qquad (4.8.7)$$

where $\sigma \in C[[0, 2\pi], E^{nN}]$. The unique solution of the corresponding IVP

$$U'(t) = AU(t) + a(t), \quad U(0) = U_0$$

is given by using (4.8.6)

$$U(t) = e^{At}U_0 + \int_0^t e^{A(t-s)}\sigma(s)ds, \quad 0 \leq t \leq 2\pi.$$

Hence if (4.8.7) is solvable, we must have

$$U_0 = U(2\pi) = e^{2A\pi}\left[U_0 + \int_0^{2\pi} e^{-As}\sigma(s)ds\right]$$

and this is possible if we assume that $[I - e^{2A\pi}]^{-1}$ exists so that we can solve for U_0, that is

$$U_0 = [I - e^{2A\pi}]^{-1}e^{2A\pi}\int_0^{2\pi} e^{-As}\sigma(s)ds.$$

4.9 Notes and Comments

The comparison theorem and the useful corollaries in terms of Lyapunov-like functions described in Section 4.2 are taken from Lakshmikantham and Leela [59]. Section 4.3 contains stability theorems parallel to the original theorems of Lyapunov and are new. The method of perturbing Lyapunov functions and the nonuniform stability results of Section 4.4 are due to Lakshmikantham and Leela [62]. The notions of boundedness and the sufficient condition for the boundedness concepts to hold in terms of Lyapunov-like functions, given in Section 4.5, are taken from Mohapatra and Zhang [76].

The description of fuzzy differential system and the corresponding comparison theorem in terms of generalized metric spaces developed in Section 4.6 are new and are modeled on the corresponding results in differential equations without fuzziness. See Lakshmikantham and Leela [59] and Bernfeld and Lakshmikantham [4]. The method of vector Lyapunov functions discussed in Section 4.7 is also new. This method is very popular and effective in applications. See Siljak [106], and Lakshmikantham, Matrosov, and Sivasundaram [64]. Section 4.8 incorporates new results on linear fuzzy systems including the variation of parameters formula and a simple criterion for periodic boundary value problems. See also Zhang et al. [120] for the solution of first-order differential equations in a special case.

Chapter 5

Miscellaneous Topics

5.1 Introduction

We initiate several interesting topics in this chapter, dealing with fuzzy dynamic equations which are yet to be investigated.

In Section 5.2 we introduce fuzzy difference equations. Since the study of difference equations has attracted many researchers, it is hoped that the investigation of fuzzy difference equations would be popular as well.

Section 5.3 initiates the development of impulsive fuzzy differential equations. Impulsive differential equations have become popular and useful recently, and therefore the typical results provided in this section would be equally attractive.

Functional differential equations or differential equations with delay are considered in Section 5.4 which is a well-known branch of differential equations. Consequently, fuzzy differential equations with delay should be an equally interesting area of research. Some typical results are incorporated in this section.

The results of Section 5.5 are new and deal with the extension of the theory of hybrid systems to fuzzy differential equations. The contents of Section 5.6 investigate the existence of fixed points of fuzzy mappings via the theory of fuzzy differential equations. Section 5.7 attempts the development of boundary value problems for fuzzy differential equations. Finally, in Section 5.8, fuzzy integral equations are discussed.

5.2   Fuzzy Difference Equations

Let N denote the natural numbers and N^+ the nonnegative natural numbers. We denote by $N_{n_0}^+$ the set

$$N_{n_0}^+ = \{n_0, n_0 + 1, n_0 + 2, ..., n_0 + k, ...\}$$

with $k \in N^+$ and $n_0 \in N$. Let us consider the fuzzy difference equation given by

$$u_{n+1} = f(n, u_n), \quad u_{n_0} = u_0, \tag{5.2.1}$$

where $f(n, u)$ is continuous in u for each n. Here $u_n, f \in E^q$ for each $n \geq n_0$, where (E^q, d) is the metric space. Since we shall be using n for difference equations, we shall employ the metric space (E^q, d) for (E^n, d) used earlier. This will avoid confusion. The possibility of obtaining the values of solutions of (5.2.1) recursively is very important and does not have a counterpart in other kinds of equations. For this reason, we sometimes reduce continuous problems to approximate difference problems. For simple fuzzy difference equations, we can find solutions in closed form. However, reducing information on the qualitative and quantitative behavior of solutions of (5.2.1) by the comparison principle is very effective as usual.

We need the following comparison principle for difference equations. See Lakshmikantham and Trigiante [69] for details.

Theorem 5.2.1. *Let $n \in N_{n_0}^+, r \geq 0$ and $g(n, r)$ be a nondecreasing function in r for each n. Suppose that for $n \geq n_0$, the inequalities*

$$y_{n+1} \leq g(n, y_n), \tag{5.2.2}$$

$$z_{n+1} \geq g(n, z_n), \tag{5.2.3}$$

hold. Then $y_{n_0} \leq z_{n_0}$ implies $y_n \leq z_n$ for all $n \geq n_0$.

Proof.   Suppose that the claim $y_n \leq z_n$ for all $n \geq n_0$ is not true. Then because of the assumption $y_{n_0} \leq z_{n_0}$, there exists a $k \in N_{n_0}^+$ such that $y_k \leq z_k$ and $y_{k+1} > z_{k+1}$. It then follows, using the monotone character of $g(n, r)$ in $r \geq 0$ and the inequalities (5.2.2) and (5.2.3), that

$$g(k, z_k) \leq z_{k+1} < y_{k+1} \leq g(k, y_k) \leq g(k, z_k).$$

This is a contradiction, which proves the claim.

The following corollaries would be useful.

Corollary 5.2.1. *Let $n \in N_{n_0}^+, k_n \geq 0$ and $y_{n+1} \leq k_n y_n + p_n$. Then*

$$y_n \leq y_{n_0} \prod_{s=n_0}^{n-1} k_s + \sum_{s=n_0}^{n-1} p_s \prod_{\tau=s+1}^{n-1} k_\tau, \quad n \geq n_0. \qquad (5.2.4)$$

Proof. Since $k_n \geq 0$ and $g(n,r) = k_n r + p_n$, the assumptions of Theorem 5.2.1 are satisfied. Take z_n as the solution of $z_{n+1} = k_n z_n + p_n$ with $z_{n_0} = y_{n_0}$ which can be computed easily. Then the result follows from Theorem 5.2.1.

Corollary 5.2.2. (Discrete Gronwall Inequality). *Let $n \in N_{n_0}^+, k_n \geq 0$ and*

$$y_{n+1} \leq y_{n_0} + \sum_{s=n_0}^{n} [k_s y_s + p_s].$$

Then

$$y_n \leq y_{n_0} \prod_{s=n_0}^{n-1} (1 + k_s) + \sum_{s=n_0}^{n-1} p_s \prod_{\tau=s+1}^{n-1} (1 + k_\tau) \qquad (5.2.5)$$

$$\leq y_{n_0} \exp\left(\sum_{s=n_0}^{n-1} k_s\right) + \sum_{s=n_0}^{n-1} p_s \exp\left(\sum_{\tau=s+1}^{n-1} k_\tau\right), \quad n \geq n_0.$$

Proof. The comparison equation is

$$z_n = z_{n_0} + \sum_{s=n_0}^{n-1} [k_s z_s + p_s], \quad z_{n_0} = y_{n_0}.$$

The solution of this equation is the expression on the right-hand side of (5.2.5). Observing $1 + k_s \leq \exp(k_s)$, we get the final expression in (5.2.5).

Let us now discuss estimating the solution of (5.2.1) in terms of solutions of the scalar difference equation

$$z_{n+1} = g(n, z_n), \quad z_{n_0} = z_0 \qquad (5.2.6)$$

where $g(n,r)$ is continuous in r for each n and nondecreasing in r for each n. We prove the following result.

Theorem 5.2.2. *Assume that $f(n,u)$ is continuous in u for each n and*

$$d[f(n,u), \hat{0}] \leq g(n, d[u, \hat{0}]), \qquad (5.2.7)$$

where $g(n,r)$ is given in (5.2.6). Then $d[u_{n_0}, \hat{0}] \leq z_{n_0}$ implies

$$d[u_{n+1}, \hat{0}] \leq z_{n+1}, \quad for \quad n \geq n_0. \qquad (5.2.8)$$

Proof.   Set $y_{n+1} = d[u_{n+1}, \hat{0}]$ so that (5.2.7) gives

$$\begin{aligned}
y_{n+1} &= d[f(n, u_n), \hat{0}] \\
&\leq g(n, d[u_n, \hat{0}] \\
&= g(n, y_n), \quad n \geq n_0.
\end{aligned}$$

Let z_{n+1} be the solution of (5.2.6), with $z_{n_0} = y_{n_0}$. Then Theorem 5.2.1 yields immediately

$$y_{n+1} \leq z_{n+1}, \quad n \geq n_0,$$

which implies (5.2.8) completing the proof.

The assumption (5.2.7) can be replaced by a weaker condition, namely,

$$d[f(n, u), \hat{0}] \leq d[u, \hat{0}] + w(n, d[u, \hat{0}]), \quad u \in E^q, \tag{5.2.9}$$

where we now set $g(n, r) = r + w(n, r)$, and assume that $g(n, r)$ is nondecreasing in r for each n. This version of Theorem 5.2.2 is more suitable because $w(n, r)$ need not be positive and hence the solutions of (5.2.6) could have better properties. This observation is also useful in extending the Lyapunov-like method for (5.2.1).

Let $V(n, u)$ be a function defined on $N_{n_0}^+ \times E^q$ which takes values in R_+. Then we have the following comparison result.

Theorem 5.2.3. *Let $V(n, u)$ defined as above satisfy*

$$\begin{aligned}
V(n+1, u_{n+1}) &\leq V(n, u_n) + w(n, V(n, u_n)) \\
&\equiv g(n, V(n, u_n)), \quad n \geq n_0.
\end{aligned}$$

Then $V(n_0, u_{n_0}) \leq z_{n_0}$ implies

$$V(n+1, u_{n+1}) \leq z_{n+1}, \quad n \geq n_0, \tag{5.2.10}$$

where $z_{n+1} = z_{n+1}(n_0, z_{n_0})$ is the solution of (5.2.6).

Proof.   Set $y_{n+1} = V(n+1, u_{n+1})$, so that $y_{n_0} = V(n_0, u_{n_0}) \leq z_{n0}$ and

$$y_{n+1} \leq y_n + w(n, y_n), \quad n \geq n_0.$$

Consequently, $g(n, r) = r + w(n, r)$. Hence by Theorem 5.2.1, we get $y_{n+1} \leq z_{n+1}$, $n \geq n_0$, where z_{n+1} is the solution of (5.2.7). This implies the stated estimate (5.2.10).

Having Theorem 5.2.3 at our disposal, it is easy to prove stability results for (5.2.1).

Theorem 5.2.4. *Let the assumptions of Theorem 5.2.3 hold. Suppose further that*

$$b(d[u,\hat{0}]) \le V(n,u) \le a(d[u,\hat{0}]),$$

where $a, b \in \mathcal{K}$, $n \in N_{n_0}^+$ *and* $u \in E^q$. *Then the stability properties of the trivial solution of (5.2.6) imply the corresponding stability properties of the trivial solution of (5.2.1).*

Proof. Suppose that the trivial solution of (5.2.6) is asymptotically stable. Then it is stable. Let $0 < \epsilon$, $n_0 \in N$ be given. Then, given $b(\epsilon) > 0$, $n_0 \in N$, there exists a $\delta_1 = \delta_1(n_0, \epsilon)$ such that

$$0 \le z_{n_0} < \delta_1 \quad \text{implies} \quad z_{n+1} < b(\epsilon), \quad n \ge n_0.$$

Choose $\delta = \delta(n_0, \epsilon)$ satisfying

$$a(\delta) < \delta_1.$$

Then Theorem 5.2.3 gives

$$V(n+1, u_{n+1}) \le z_{n+1}, \quad n \ge n_0,$$

which shows that

$$b(d[u_{n+1}, \hat{0}]) \le V(n+1, u_{n+1}) \le z_{n+1}, \quad n \ge n_0.$$

Choose $z_{n_0} = V(n_0, u_{n_0})$ so that we have

$$z_{n_0} \le a(d[u_{n_0}, \hat{0}]) \le a(\delta) < \delta_1.$$

We then get

$$b(d[u_{n+1}, \hat{0}]) < b(\epsilon), \quad n \ge n_0,$$

which implies the stability of the trivial solution of (5.2.1).

For asymptotic stability, we observe that

$$b(d[u_{n+1}, \hat{0}]) \le V(n+1, u_{n+1}) \le z_{n+1}, \quad n \ge n_0.$$

Since $z_{n+1} \to 0$ as $n \to \infty$, we get $d[u_{n+1}, \hat{0}] \to 0$ as $n \to \infty$. The proof is complete.

As an example, take $g(n, r) = a_n r$ where $a_n \in R$. Then the solution of

$$z_{n+1} = a_n z_n, \quad z_{n_0} = z_0,$$

is given by

$$z_n = z_0 \prod_{i=n_0}^{n-1} a_i.$$

We have the following two cases:

(a) If

$$\left| \prod_{i=n_0}^{n-1} a_i \right| \leq M(n_0),$$

then $|z_n| \leq |z_0| M(n_0)$ and therefore it is sufficient to take $\delta(\epsilon, n_0) = \frac{\epsilon}{M(n_0)}$, to get stability.

(b) If

$$\lim_{n \to \infty} \left| \prod_{i=n_0}^{n-1} a_i \right| = 0,$$

then asymptotic stability results.

Consequently, Theorem 5.2.3 yields the corresponding stability properties of the trivial solution of (5.2.1).

5.3   Impulsive Fuzzy Differential Equations

Let PC denote the class of piecewise continuous functions from R_+ to R with discontinuities of the first kind only at $t = t_k$, $k = 1, 2, \dots$. We need the following known result.

Theorem 5.3.1. *Assume that*

(A_0) *The sequence $\{t_k\}$ satisfies $0 \leq t_0 < t_1 < t_2, \cdots$ with $\lim_{k \to \infty} t_k = \infty$;*

(A_1) *$m \in PC^1[R_+, R]$ and $m(t)$ is left continuous at $t_k, k = 1, 2, \dots$;*

(A_2) *for $k = 1, 2, \dots, t \geq t_0$,*

$$m'(t) \leq g(t, m(t)), \quad t \neq t_k, \quad m(t_0) \leq w_0,$$

$$m(t_k^+) \leq \psi_k(m(t_k)), \tag{5.3.1}$$

where $g \in C[R_+ \times R, R], \psi_k : R \to R, \psi_k(w)$ is nondecreasing in w;

(A_3) *$r(t) = r(t, t_0, w_0)$ is the maximal solution of*

$$w' = g(t, w), \quad t \neq t_k, \quad w(t_0) = w_0,$$

$$w(t_k^+) = \psi_k(w(t_k)), \quad t_k > t_0 \geq 0, \tag{5.3.2}$$

existing on $[t_0, \infty)$.

Then $m(t) \leq r(t), t \geq t_0$.

Proof.   For $t \in [t_0, t_1]$, we have by the classical comparison theorem $m(t) \leq r(t)$. Hence, using the facts that $\psi_1(w)$ is nondecreasing in w and $m(t_1) \leq r(t_1)$, we obtain

$$m(t_1^+) \leq \psi_1(m(t_1)) \leq \psi_1(r(t_1)) = w_1^+.$$

Now, for $t_1 < t \leq t_2$, it follows, using again the classical comparison theorem, that $m(t) \leq r(t)$, where $r(t) = r(t, t_1, w_1^+)$ is the maximal solution of (5.3.2) on the interval $t_1 \leq t \leq t_2$. Moreover, as before, we get

$$m(t_2^+) \leq \psi_2(m(t_2)) \leq \psi_2(r(t_2)) = w_2^+.$$

Repeating the arguments, we finally arrive at the desired result and the proof is complete. See Lakshmikantham, Bainov and Simeonov [58] for details.

Let us consider now the impulsive fuzzy differential equation

$$u' = f(t, u), \quad t \neq t_k,$$

$$u(t_k^+) = u(t_k) + I_k(u(t_k)), \quad u(t_0) = u_0, \tag{5.3.3}$$

where (A_0) holds and $f : R_+ \times E^n \rightarrow E^n$, $I_k : E^n \rightarrow E^n$, f is continuous in $(t_{k-1}, t_k] \times E^n$ and for each $u \in E^n$, $\lim f(t, v) = f(t_k^+, u)$ exists as $(t, v) \rightarrow (t_k^+, u)$. We shall obtain the bounds for the solution of (5.3.3). If the assumptions of Theorem 3.2.1 hold for each $(t_{k-1}, t_k] \times E^n$, then clearly there exists a unique solution $u(t)$ of (5.3.3) in each interval $[t_{k-1}, t_k]$. As a result, employing the impulsive condition in (5.3.3) at each t_k, we can define the solution $u(t)$ on the entire interval $[t_0, \infty)$ as in the foregoing discussion of the proof of Theorem 5.3.1.

Theorem 5.3.2. *Assume that $f \in C[R_+ \times E^n, E^n]$ and*

$$\limsup_{h \rightarrow 0+} \frac{1}{h} [d[u + hf(t, u), v + hf(t, v)] - d[u, v]]$$

$$\leq g(t, d[u, v]), \quad t \in R_+, \quad u, v \in E^n, \quad t \neq t_k,$$

where $g \in C[R_+ \times R_+, R]$. Suppose that

$$d[u + I_k(u), v + I_k(v)] \leq \psi_k(d[u, v])$$

where $\psi_k : R_+ \rightarrow R_+, \psi_k(w)$ is nondecreasing in w. The maximal solution $r(t) = r(t, t_0, w_0)$ of (5.3.2) exists for $t \geq t_0$. Then

$$d[u(t), v(t)] \leq r(t), \quad t \geq t_0,$$

where $u(t), v(t)$ are the solutions of (5.3.3) existing on $[t_0, \infty)$.

Proof.   Proceeding as in the proof of Theorem 3.4.2, we find that for $t \neq t_k$,

$$
\begin{aligned}
m(t+h) - m(t) &= d[u(t+h), v(t+h)] - d[u(t), v(t)] \\
&\leq d[u(t+h, u(t) + hf(t, u(t)) + d[v(t) + hf(t, v(t)), v(t+h)] \\
&\quad + d[hf(t, u(t)), hf(t, v(t))] - d[u(t), v(t)].
\end{aligned}
$$

Hence

$$
D^+ m(t) = \lim \sup_{h \to 0^+} \frac{1}{h}[m(t+h) - m(t)]
$$

$$
\begin{aligned}
&\leq \lim \sup_{h \to 0^+} \frac{1}{h}[d(u(t) + hf(t, u(t)), \ v(t) + hf(t, v(t))] - d[h(t), v(t)] \\
&\quad + \lim \sup_{h \to 0^+} d\left[\frac{u(t+h) - u(t)}{h}, f(t, u(t))\right] \\
&\quad + \lim \sup_{h \to 0^+} d[f(t, v(t)), \ v(t) + hf(t, v(t))], \ t \neq t_k \\
&\leq g(t, d[u(t), \ v(t)]) \\
&= g(t, m(t)), \quad t \neq t_k.
\end{aligned}
$$

Also,

$$
\begin{aligned}
m(t_k^+) &= d[u(t_k^+), v(t_k^+)] \\
&= d[u(t_k) + I_k(u(t_k)), v(t_k) + I_k(v(t_k))] \\
&\leq \psi_k(d[u(t_k), v(t_k)] \\
&= \psi_k(m(t_k)), \quad t \neq t_k.
\end{aligned}
$$

We therefore obtain from Theorem 5.3.1, the stated result, namely,

$$
d[u(t), v(t)] \leq r(t), \quad t \geq t_0,
$$

where $r(t) = r(t, t_0, w_0)$ is the maximal solution of (5.3.2) provided $d[u_0, v_0] \leq w_0$, completing the proof.

We shall next extend a typical result in Lyapunov-like theory.

Let $V : R_+ \times E^n \to R_+$. Then V is said to belong to class V_0 if

(i) V is continuous in $(t_{k-1}, t_k] \times E^n$ and for each $u \in E^n, k = 1, 2, ...,$
$\lim_{(t,v) \to (t_k^+, u)} V(t, v) = V(t_k^+, u)$ exists;

(ii) V satisfies $|V(t, u) - V(t, v)| \leq L d[u, v], L \geq 0$.

For $(t, u) \in (t_{k-1}, t_k] \times E^n$, we define

$$D^+V(t, u) = \lim \sup_{h \to 0+} \frac{1}{h}[V(t + u), u + hf(t, u)) - V(t, u)].$$

Then we can prove the following comparison theorem.

Theorem 5.3.3. *Let* $V : R_+ \times E^n \to R_+$ *and* $V \in V_0$. *Suppose that*

$$D^+V(t, u) \leq g(t, V(t, u)), \quad t \neq t_k, \tag{5.3.4}$$

$$V(t, u + I_k(u)) \leq \psi_k(V(t, u)), \quad t = t_k, \tag{5.3.5}$$

where $g : R_+^2 \to R$ *is continuous in* $(t_{k-1}, t_k] \times R_+$ *and for each* $w \in R_+$, $\lim_{(t,z) \to (t_k^+, w)} g(t, z) = g(t_k^+, w)$ *exists,* $\psi_k : R_+ \to R$ *is nondecreasing.*
Let $r(t)$ *be the maximal solution of the scalar impulsive differential equation* (5.3.2) *existing for* $t \geq t_0$. *Then* $V(t_0^+, u_0) \leq w_0$ *implies* $V(t, u(t)) \leq r(t), t \geq t_0$.

Proof. Let $u(t) = u(t, t_0, u_0)$ be any solution of (5.3.3) existing on $t \geq t_0$ such that $V(t_0^+, u_0) \leq w_0$. Define $m(t) = V(t, u(t))$ for $t \neq t_k$. Then proceeding as in the proof of Theorem 4.2.1, we arrive at the differential inequality

$$D^+m(t) \leq g(t, m(t)), \quad t \neq t_k.$$

From (5.3.5), we get for $t = t_k$

$$
\begin{aligned}
m(t_k^+) &= V(t_k^+, u(t_k^+)) \\
&= V(t_k^+, u(t_k) + I_k(u(t_k))) \\
&\leq \psi_k(V(t_k, u(t_k))) \\
&= \psi_k(m(t_k)).
\end{aligned}
$$

Hence by Theorem 5.3.1, $m(t) \leq r(t), t \geq t_0$, which proves the claim of Theorem 5.3.3.

Some special cases of $g(t, w)$ and $\psi_k(w)$ which are instructive and useful are given below as a corollary.

Corollary 5.3.1. *In Theorem 5.3.3, suppose that*

(1) $g(t, w) = 0, \psi_k(w) = w$ *for all* k, *then* $V(t, u(t))$ *is nondecreasing in* t *and* $V(t, u(t) \leq V(t_0^+, u_0), t \geq t_0$;

(2) $g(t, w) \equiv 0, \psi_k(w) = d_k w, d_k \geq 0$ for all k, then

$$V(t, u(t)) \leq V(t_0^+, u_0) \prod_{t_0 < t_k < t} d_k, \quad t \geq t_0;$$

(3) $g(t, w) = -\alpha w, \alpha > 0, \psi_k(w) = d_k w, d_k \geq 0$ for all k, then

$$V(t, u(t)) \leq \left[V(t_0^+, u_0) \prod_{t_0 < t_k < t} d_k \right] e^{-\alpha(t - t_0)}, \quad t \geq t_0;$$

(4) $g(t, w) = \lambda'(t) w, \psi_k(w) = d_k w, d_k \geq 0$ for all $k, \lambda \in C^1[R_+, R_+]$, then

$$V(t, u(t)) \leq \left[V(t_0^+, u_0) \prod_{t_0 < t_k < t} d_k \right] \exp[\lambda(t) - \lambda(t_0)] \quad t \geq t_0.$$

We shall give a typical result on stability criteria.

Theorem 5.3.4. *Assume that*

(i) $V : R_+ \times S(\rho) \to R_+, V \in V_0, S(\rho) = [u \in E^n : d[u, \hat{0}] < \rho]$ *and* $D^+ V(t, u) \leq g(t, V(t, u)), t \neq t_k$, *where* $g : R_+^2 \to R, g(t, 0) \equiv 0$ *and* g *satisfies the assumption given in Theorem 5.3.3;*

(ii) *there exists a* $\rho_0 > 0$ *such that* $u \in S(\rho_0)$ *implies* $u + I_k(u) \in S(\rho)$ *for all* k, *and*

$$V(t, u + I_k(u)) \leq \psi_k(V(t, u)), \quad t = t_k, \quad u \in S(\rho_0),$$

where $\psi_k : R_+ \to R_+$ *is nondecreasing;*

(iii) $b(d[u, \hat{0}]) \leq V(t, u) \leq a(d[u, \hat{0}]), (t, u) \in R_+ \times S(\rho)$, *where* $a, b \in \mathcal{K}$.

Then the stability properties of the trivial solution of (5.3.2) *imply the corresponding stability properties of the trivial solution of* (5.3.3).

Proof.   We shall give the proof of stability only. Let $0 < \epsilon < \rho^* = \min(\rho, \rho_0), t_0 \in R_+$ be given. Suppose that the trivial solution of (5.3.2) is stable. Then given $b(\epsilon) > 0$ and $t_0 \in R_+$, there exists a $\delta_1 = \delta_1(t_0, \epsilon) > 0$ such that

$$0 \leq w_0 < \delta_1 \quad \text{implies} \quad w(t, t_0, w_0) < b(\epsilon), \quad t \geq t_0,$$

where $w(t, t_0, w_0)$ is any solution of (5.3.2). Let $w_0 = a(d[u_0, \hat{0}])$ and choose a $\delta = \delta(\epsilon)$ such that $a(\delta) < \delta_1$. With this δ, we claim that if

$$d[u_0, \hat{0}] < \delta \quad \text{then} \quad d[u(t), \hat{0}] < \epsilon, \quad t \geq t_0.$$

If this is not true, there would exist a solution $u(t) = u(t, t_0, u_0)$ of (5.3.3) with $d[u_0, \hat{0}] < \delta$ and a $t^* > t_0$ such that $t_k < t^* \leq t_{k+1}$ for some k^*, satisfying

$$\epsilon \leq d[u(t^*), \hat{0}] \quad \text{and} \quad d[u(t), \hat{0}] < \epsilon, \quad \text{for} \quad t_0 \leq t \leq t_k.$$

Since $0 < \epsilon < \rho_0$, condition (ii) shows that

$$d[u(t_k^+), \hat{0}] = d[u(t_k) + I_k(u(t_k)), \hat{0}] < \rho \quad \text{and} \quad d[u(t_k), \hat{0}] < \epsilon.$$

Hence we can find a t^0 such that $t_k < t^0 \leq t^*$ satisfying $\epsilon \leq d[u(t^0), \hat{0}] < \rho$. Now setting $m(t) = V(t, u(t))$ for $t_0 \leq t \leq t^0$ and using (i) and (ii) we get by Theorem 5.3.3, the estimate

$$V(t, u(t)) \leq r(t, t_0, a(d[u_0, \hat{0}])), \quad t_0 \leq t \leq t^0,$$

where $r(t, t_0, w_0)$ is the maximal solution of (5.3.2). We are then led to a contradiction because of (iii):

$$b(\epsilon) \leq b(d[u(t^0), \hat{0}]) \leq V(t^0, u(t^0)) \leq r(t^0, t_0, a(d[u_0, \hat{0}])) < b(\epsilon),$$

which proves that the trivial solution of (5.3.3) is stable.

As an example, consider the special case given in Corollary 5.3.1, namely, $g(t, w) = \lambda'(t)w$, $\psi_k(w) = d_k w$, $d_k \geq 0$ for all k and $\lambda \in C^1[R_+j, R_+]$ with $\lambda'(t) \geq 0$. If λ satisfies

$$\lambda(t_{k+1}) + \ln d_k \leq \lambda(t_k) \quad \text{for all} \quad k, \tag{5.3.6}$$

then the trivial solution of (5.3.2) is stable and hence the trivial solution of (5.3.3) is also stable. This follows because the solution $w(t, t_0, w_0)$ of (5.3.2) in this case would be

$$w(t, t_0, w_0) = w_0 \prod_{t_0 < t_k < t} d_k \exp[\lambda(t) - \lambda(t_0)], \quad t \geq t_0.$$

Since $\lambda(t)$ is nondecreasing, it follows from (5.3.6) that

$$w(t, t_0, w_0) \leq w_0 \exp[\lambda(t_1) - \lambda(t_0)], \quad t \geq t_0,$$

provided $0 < t_0 < t_1$. Hence choosing $\delta = \frac{\epsilon}{2} \exp[\lambda(t_0) - \lambda(t_1)]$ stability follows.

5.4   Fuzzy Differential Equations with Delay

Given any $\tau > 0$, we let $C_0 = C[[-\tau, 0], E^n]$. For any element $\varphi \in C_0$, define
the metric $H[\varphi, \psi] = \max_{-\tau \leq s \leq 0} d[\varphi(s), \psi(s)]$. Suppose that $u \in C[J_0, E^n]$
where $J_0 = [t_0 - \tau, t_0 + a]$, $a > 0$. For any $t \geq t_0$, $t \in J_0$, we let u_t denote a
translation of the restriction of u to the interval $[t-\tau, t]$; specifically, $u_t \in C_0$
defined by

$$u_t(s) = u(t + s), \quad -\tau \leq s \leq 0.$$

In other words, the graph of u_t is the graph of u on $[t-\tau, t]$ shifted to $[-\tau, 0]$.
With this notation, we consider the fuzzy differential equation with finite
delay

$$u'(t) = f(t, u_t), \quad u_{t_0} = \varphi_0 \in C_0, \tag{5.4.1}$$

where $f, u \in E^n$ and $f \in C[J \times C_0, E^n]$, $J = [t_0, t_0 + a]$. We can then prove
the following existence result.

Theorem 5.4.1. *Assume that*

$$d[f(t, \varphi), f(t, \psi)] \leq kH[\varphi, \psi], \quad k > 0, \tag{5.4.2}$$

for $t \in J$ and $\varphi, \psi \in C_0$. Then the IVP (5.4.1) possesses a unique solution
$u(t)$ on J_0.

Proof.   Consider the space of functions $u \in C[J_0, E^n]$ such that $u(t) =$
$\varphi_0(t)$, $t_0 - \tau \leq t \leq t_0$ and $u \in C[J, E^n]$ with $u(t_0) = \varphi_0(0)$ and $\varphi_0(t) \in E^n$
for $-\tau \leq t \leq 0$. Define the metric on $C[J_0, E^n]$ by

$$\rho[u, v] = \max_{t_0 - \tau \leq t \leq t_0 + a} d[u(t), v(t)]e^{-\lambda t}, \quad \lambda > 0. \tag{5.4.3}$$

Next we define the operator T on $C[J_0, E^n]$ by

$$\left.\begin{array}{ll} Tu(t) &= \varphi_0(t), \quad t_0 - \tau \leq t \leq t_0, \\ Tu(t) &= \varphi_0(0) + \int_{t_0}^{t} f(s, u_s)ds, \quad t \in J. \end{array}\right\} \tag{5.4.4}$$

We find that $d[Tu(t_0 + s), Tv(t_0 + s)] = 0, -\tau \le s \le 0$, and

$$d[Tu(t), Tv(t)] \le d\left[\int_{t_0}^t f(s, u_s)ds, \int_{t_0}^t f(s, v_s)ds\right]$$

$$\le \int_{t_0}^t d[f(s, u_\xi), f(s, v_\xi)]d\xi$$

$$\le k\int_{t_0}^t d[u_\xi, v_\xi]d\xi$$

$$= k\int_{t_0}^t d[u(\xi + s), v(\xi + s)]d\xi$$

$$= k\int_{t_0-\tau}^t d[u(\sigma), v(\sigma)]d\sigma.$$

Hence we have on J_0,

$$e^{-\lambda t}d[Tu(t), Tv(t)] \le ke^{-\lambda t}\int_{t_0-\tau}^t d[u(\sigma), v(\sigma)]e^{-\lambda\sigma}e^{\lambda\sigma}d\sigma$$

$$\le ke^{-\lambda t}\rho[u, v]\int_{t_0-\tau}^t e^{\lambda\sigma}d\sigma$$

$$\le \frac{k}{\lambda}\rho[u, v].$$

Choosing $\lambda = 2k$, we arrive at

$$\rho[Tu, Tv] \le \frac{1}{2}\rho[u, v],$$

showing that the operator T on $C[J_0, E^n]$ is a contraction. We therefore obtain a unique fixed point $u \in C[J_0, E^n]$ of T by the contraction mapping principle, and consequently this $u = u(t)$ is the unique solution of the IVP (5.4.1) on J_0. The proof is complete.

We need the following known result.

Theorem 5.4.2. *Let $m \in C[[t_0 - \tau, \infty], R_+]$ such that*

$$D^+m(t) \le g(t, |m_t|_0), \quad t > t_0,$$

where $g \in C[R_+^2, R], |m_t|_0 = \max_{-\tau \le s \le 0}|m(t + s)|$. Assume that $r(t) = r(t, t_0, w_0)$ is the maximal solution of the scalar differential equation

$$w' = g(t, w), \quad w(t_0) = w_0 \ge 0, \tag{5.4.5}$$

existing on $[t_0, \infty)$. *Then* $|m_{t_0}|_0 \leq w_0$ *implies that*

$$m(t) \leq r(t), \quad t \geq t_0.$$

See Lakshmikantham and Leela [60] *for the proof and other details.*

We can now prove the following result relative to the solutions of (5.4.1).

Theorem 5.4.3. *Assume that* $f \in C[R_+ \times C_0, E^n]$ *and*

$$d[f(t, \varphi), f(t, \psi)] \leq g(t, H[\varphi, \psi]), \qquad (5.4.6)$$

for $t \in R_+, \varphi, \psi \in C_0$, *where* $g \in C[R_+ \times R_+, R_+]$. *Let* $r(t) = r(t, t_0, w_0)$ *be the maximal solution of* (5.4.5) *existing for* $t \geq t_0$. *Then* $H[\varphi_0, \psi_0] \leq w_0$ *implies*

$$d[u(t), v(t)] \leq r(t), \quad t \geq t_0, \qquad (5.4.7)$$

where $u(t) = u(t_0, \varphi_0)(t)$, $v(t) = v(t_0, \psi_0)(t)$ *are the solutions of* (5.4.1).

Proof.   Setting $m(t) = d[u(t), v(t)]$, where $u(t) = u(t_0, \varphi_0)(t)$, $v(t) = v(t_0, \psi_0)(t)$ are the solutions of (5.4.1) and proceeding as in Theorem 3.4.2, we find

$$
\begin{aligned}
D^+ m(t) &= \limsup_{h \to 0} \frac{1}{h}[m(t+h) - m(t)] \\
&\leq \limsup_{h \to 0+} d\left[\frac{u(t+h) - u(t)}{h}, f(t, u_t)\right] \\
&\quad + \limsup_{h \to 0+} d\left[f(t, v_t), \frac{v(t+h) - v(t)}{h}\right] \\
&\quad d[f(t, u_t, f(t, v_t)], \quad t_0 < t < \infty.
\end{aligned}
$$

Then condition (5.4.6) implies that

$$D^+ m(t) \leq g(t, H[u_t, v_t]) = g(t, |m_t|_0), \quad t_0 < t < \infty.$$

We also have $|m_{t_0}| = H[\varphi_0, \psi_0] \leq w_0$ and hence by Theorem 5.4.2 we get the desired estimate

$$d[u(t), v(t) \leq r(t), \quad t \geq t_0.$$

Let us next prove a global existence result.

Theorem 5.4.4. *Let $f \in C[R_+ \times C_0, E^n]$ and for $(t, \varphi) \in R_+ \times C_0$,*

$$d[f(t, \varphi), \hat{0}] \leq g(t, H[\varphi, \hat{0}])$$

where $g \in C[R_+^2, R_+]$, $g(t, w)$ is nondecreasing in w for each $t \in R$. Assume that the solutions $w(t, t_0, w_0)$ of (5.4.5) exist for $t \geq t_0$ and f is smooth enough to assure local existence. Then the largest interval of existence of any solution $u(t_0, \varphi_0)$ of (5.4.1) is $[t_0, \infty)$.

Proof. Let $u(t_0, \varphi_0)(t)$ be a solution of (5.4.1) existing on some interval $[t_0 - \tau, \beta)$, where $t_0 < \beta < \infty$. Assume that β cannot be increased. Define for $[t_0 - \tau, \beta)$, $m(t) = d[u(t_0, \varphi_0)(t), \hat{0}]$ so that $m_t = d[u_t(t_0, \varphi_0), \hat{0}]$. Then using the arguments of Theorem 3.4.2 and Theorem 5.4.3, we get the differential inequalities

$$D^+ m(t) \leq g(t, |m_t|_0), \quad t_0 \leq t < \beta.$$

Choosing $|m_{t_0}|_0 = |\varphi_0|_0 \leq w_0$, we get

$$d[u(t_0, \varphi_0)(t)\,\hat{0}] \leq r(t, t_0, w_0), \quad t_0 \leq t < \beta. \tag{5.4.8}$$

Since $g(t, w) \geq 0$ and as a result, $r(t, t_0, w_0)$ is nondecreasing in t, we have

$$d[u_t(t_0, \varphi_0), \hat{0}] \leq r(t, t_0, w_0), \quad t_0 \leq t < \beta. \tag{5.4.9}$$

For any t_1, t_2 such that $t_0 < t_1 < t_2 < \beta$, one gets

$$d[u(t_0, \varphi_0)(t_1), u(t_0, \varphi_0)(t_2)] \leq \int_{t_1}^{t_2} g(s, H[u_s(t_0, \varphi_0), \hat{0}]) ds$$

which, in view of (5.4.9) and the monotonicity of $g(t, w)$ in w, implies

$$\begin{aligned} d[u(t_0, \varphi_0)(t_1), u(t_0, \varphi_0)(t_2)] &\leq \int_{t1}^{t2} g(s, r(s, t_0, w_0)) ds \\ &= r(t_2, t_0, w_0) - r(t_1, t_0, w_0). \end{aligned} \tag{5.4.10}$$

Letting $t_1, t_2 \to \beta$, the foregoing relation shows that $\lim_{t \to \beta-} u(t_0, \varphi_0)(t)$ exists, because of Cauchy's criteria for convergence. We now define $u(t_0, \varphi_0)(\beta) = \lim_{t \to \beta-} u(t_0, \varphi_0)(t)$ and consider $\psi_0 = u_\beta(t_0, \varphi_0)$ as the new initial function at $t = \beta$. The assumption on local existence implies that there exists a solution $u(\beta, \psi_0)$ of (5.4.1) on $[\beta - \tau, \beta + \alpha], \alpha > 0$. This means that the solution $u(t_0, \varphi_0)(t)$ can be continued beyond β, which is contrary to our assumption that the value of β cannot be increased. Hence the stated result follows.

Corollary 5.4.1. *If, in addition to the assumptions of Theorem 5.4.4, we suppose that all the solutions $w(t, t_0, w_0)$ of (5.4.5) are bounded on $[t_0, \infty)$, then every solution $u(t_0, \varphi_0)(t)$ of (5.4.1) tends to a finite limit ξ as $t \to \infty$.*

Proof.   The assumption of boundedness of solutions $w(t, t_0, w_0)$ on $[t_0, \infty)$ implies that $\lim_{t\to\infty} w(t, t_0, w_0) = w^*$ (finite limit). This means that given an $\epsilon > 0$, it is possible to find a $t_1 > 0$ sufficiently large such that

$$0 \leq r(t_1, t_0, w_0) - r(t_1, t_0, w_0) < \epsilon, \quad t \geq t_1.$$

Consequently, using (5.4.10), which is now valid for all $t \geq t_0$, we get $d[u(t_0, \varphi_0)(t), u(t_0, \varphi_0)(t_1)] < \epsilon, t > t_1$, which proves $\lim_{t\to\infty} u(t_0, \varphi_0)(t) = \xi$.

5.5   Hybrid Fuzzy Differential Equations

The problem of stabilizing a continuous plant governed by differential equation through the interaction with a discrete time controller has recently been investigated. This study leads to the consideration of hybrid systems. In this section, we shall extend this approach to fuzzy differential equations.

Consider the hybrid fuzzy differential system

$$u'(t) = f(t, u(t), \lambda_k(z)), \quad u(t_k) = z, \tag{5.5.1}$$

on $[t_k, t_{k+1}]$ for any fixed $z \in E^n$, $k = 0, 1, 2, ...$, where $f \in C[R_+ \times E^n \times E^n, E^n]$, and $\lambda_k \in C[E^n, E^n]$. Here we assume that $0 \leq t_0 < t_1 < t_2 < ...$ are such that $t_k \to \infty$ as $k \to \infty$ and the existence and uniqueness of solutions of the hybrid system hold on each $[t_k, t_{k+1}]$. To be specific, the system would look like

$$u'(t) = \begin{cases} u_0'(t) &= f(t, u_0(t), \lambda_0(u_0)), \quad u_0(t_0) = u_0, \quad t_0 \leq t \leq t_1, \\ u_1'(t) &= f(t, u_1(t), \lambda_1(u_1)), \quad u_1(t_1) = u_1, \quad t_1 \leq t \leq t_2, \\ &\cdots \\ &\cdots \\ u_k'(t) &= f(t, u_k(t), \lambda_k(u_k)), \quad u_k(t_k) = u_k, \quad t_1 \leq t \leq t_{k+1}, \\ &\cdots \\ &\cdots \end{cases}$$

By the solution of (5.5.1), we therefore mean the following function

$$u(t) = u(t, t_0, u_0) = \begin{cases} u_0(t), & t_0 < t \le t_1, \\ u_1(t), & t_1 < t \le t_2 \\ \dots \\ \dots \\ u_k(t), & t_k < t \le t_{k+1} \\ \dots \\ \dots \end{cases}$$

We note that the solutions of (5.5.1) are piecewise differentiable in each interval for $t \in [t_k, t_{k+1}]$ for any fixed $u_k \in E^n$ and $k = 0, 1, 2, \dots$.

Let $V \in C[E^n, R_+]$. For $t \in (t_k, t_{k+1}]$, $u, z \in E^n$, we define

$$D^+V(u, z) = \lim_{h \to 0+} \sup \frac{1}{h}[V(u + hf(t, u, \lambda_k(z))) - V(u)].$$

We can then prove the following comparison theorem in terms of the Lyapunov -like function V.

Theorem 5.5.1. *Assume that*

(i) $V \in C[E^n, R_+]$, $V(u)$ *satisfies* $|V(u) - V(v)| \le Ld[u, v]$, $L > 0$ *for* $u, v \in E^n$;

(ii) $D^+V(u, z) \le g(t, V(u), \sigma_k(V(z)))$, $t \in (t_k, t_{k+1}]$, *where* $g \in C[R_+^3, R]$, $\sigma_k \in C[R_+, R_+]$, $u, z \in E^n$, $k = 0, 1, 2, \dots$;

(iii) *the maximal solution* $r(t) = r(t, t_0, w_0)$ *of the hybrid scalar differential equation*

$$\left. \begin{array}{ll} w' & = g(t, w(t), \sigma_k(w_k)), \quad t \in (t_k, t_{k+1}] \\ w(t_k) & = w_k, \quad k = 0, 1, 2, \dots, \end{array} \right\} \tag{5.5.2}$$

exists on $[t_0, \infty)$.

Then any solution $u(t) = u(t, t_0, u_0)$ *of (5.5.1) such that* $V(u_0) \le w_0$ *satisfies the estimate*

$$V(u(t)) \le r(t), \quad t \ge t_0.$$

Proof.   Let $u(t)$ be any solution of (5.5.1) existing on $[t_0, \infty)$ and set $m(t) = V(u(t))$. Then using (i) and (ii) and proceeding as in the proof of Theorem 4.2.1, we get the differential inequality

$$D^+m(t) \le g(t, m(t), \sigma_k(m_k)) \quad \text{for} \quad t_k < t \le t_{k+1},$$

where $m_k = V(u(t_k))$. For $t \in [t_0, t_1]$, since $m(t_0) = V(u_0) \le w_0$, we obtain by Theorem 4.2.1,

$$V(u_0(t)) \le r_0(t, t_0, w_0), \quad t_0 \le t \le t_1,$$

where $r_0(t) = r_0(t, t_0, w_0)$ is the maximal solution of

$$w_0' = g(t, w_0, \sigma_0(w_0)), \quad w_0(t_0) = w_0 \ge 0, \quad t_0 \le t \le t_1,$$

and $u_0(t)$ is the solution of

$$u_0'(t) = f(t, u_0(t), \lambda_0(u_0)), \quad u(t_0) = u_0, \quad t_0 \le t \le t_1.$$

Similarly, for $t \in [t_1, t_2]$, it follows that

$$V(u_1(t)) \le r_1(t, t_1, w_1), \quad t_1 \le t \le t_2,$$

where $w_1 = r_0(t_1, t_0, w_0), r_1(t, t_1, w_1)$ is the maximal solution of

$$w_1'(t) = g(t, w_1(t, \sigma_1(w_1)) \quad w_1(t_1) = w_1 \ge 0, \quad t_1 \le t \le t_2,$$

and $u_1(t)$ is the solution of

$$u_1'(t) = f(t, u, (t), \lambda_1(u_1)), \quad u_1(t_1) = u_1, \quad t_1 \le t \le t_2.$$

Proceeding similarly, we can obtain

$$V(u_k(t)) \le r_k(t, t_k, w_k), \quad t_k \le t \le t_{k+1},$$

where $u_k(t)$ is the solution of

$$u_k'(t) = f(t, u_k(t), \lambda_k(u_k)), \quad u_k(t_k) = u_k, \quad t_k \le t \le t_{k+1},$$

and $r_k(t, t_k, w_k)$ is the maximal solution of

$$w_k'(t) = g(t, w_k(t), \sigma_k(w_k)), \quad w_k(t_k) = w_k, \quad t_k \le t \le t_{k+1},$$

where $w_k = r_{k-1}(t_k, t_{k-1}, r_{k-2}(t_{k-1}, t_{k-2}, w_{k-1}))$. Thus defining $r(t, t_0, w_0)$ as the maximal solution of the comparison hybrid system (5.5.2) as

$$r(t, t_0, w_0) = \begin{cases} r_0 & (t, t_0, w_0), \quad t_0 \le t \le t_1, \\ r_1 & (t, t_1, w_1), \quad t_1 \le t \le t_2, \\ \quad \vdots \\ r_k & (t, t_k, w_k), \quad t_k < t \le t_{k+1} \\ \quad \cdots \\ \quad \cdots \end{cases}$$

and taking $w_0 = V(u_0)$, we obtain the desired estimate

$$V(u(t)) \leq r(t), \quad t \geq t_0.$$

The proof is therefore complete.

Once we have the comparison result, it is not difficult to discuss stability criteria of the hybrid fuzzy differential equation (5.5.1). To avoid monotony, we omit such consideration.

5.6 Fixed Points of Fuzzy Mappings

In this section we shall utilize the theory of fuzzy differential equations combined with the contraction mapping principle, to obtain fixed points of fuzzy mappings. For this purpose, let us consider the autonomous fuzzy differential equation

$$u' = f(u), \quad u(0) = u_0, \tag{5.6.1}$$

where $f \in C[E^n, E^n]$. We can prove the following global existence result.

Theorem 5.6.1. *Assume that*

(i) $\limsup_{h \to 0^+} \frac{1}{h}[d[u + hf(u), v + hf(v)] - d[u, v]] \leq -\beta d[u, v]$, $\beta > 0$, u, $v \in E^n$;

(ii) $d[f(u), \hat{0}] \leq M$ *whenever* $d[u, \hat{0}] \leq L$;

(iii) *for each* $u_0 \in E^n$, *there exists a solution locally on* $[0, a]$.

Then for each $u_0 \in E^n$, *there is a unique solution* $u(t, u_0)$ *existing on* $[0, \infty)$.

Proof. By (iii), there is a solution $u(t, x_0)$ on $[0, a]$ for some $x_0 \in E^n$. Suppose that $u(t, x_0)$ for some $x_0 \in E^n$ is not continuable, that is, it exists only on $[0, \eta), 0 < \eta < \infty$.

Set $m(t) = d[u(t), u_0], t \in [0, \eta)$. Let $u(t + h) = u(t) + z(t)$, where $z(t)$ is the H-difference.

Then we have

$$
\begin{aligned}
m(t + h) &= d[u(t + h, u_0] \\
&= d[u(t + h), u_0] - d[u(t) + hf(u), u_0] + d[u(t) + hf(u), u_0] \\
&= d[u + z, u + hf] + d[u(t) + hf(u), u_0] \\
&= d[z, hf] + d[u(t) + hf(u), u_0] \\
&= d[u(t + h) - u(t), hf(u)] + d[u(t) + hf(u), u_0].
\end{aligned}
$$

Hence

$$\frac{m(t+h) - m(t)}{h} = \frac{1}{h}[d[u(t+h) - u(t), hf(u)] + d[u(t) + hf(u), u_0] - d[u, u_0]].$$

But

$$d[u + hf(u), u_0] \leq d[u + hf(u), u_0 + hf(u_0)] + d[u_0 + hf(u_0), u_0]$$

$$\leq d[u + hf(u), u_0 + hf(u_0)] + d[hf(u_0), \hat{0}].$$

Thus

$$\begin{aligned}
\frac{m(t+h) - m(t)}{h} \quad &\leq \quad \frac{1}{h}[du(t+h) - u(t), hf(u)] \\
&+ \quad d[u + hf(u), u_0 + hf(u_0)] \\
&- \quad d[u, u_0] + d[hf(u_0), \hat{0}],
\end{aligned}$$

which implies

$$\begin{aligned}
D^+ m(t) \quad &\leq \quad \lim_{h \to 0} d\left[\frac{u(t+h) - u(t)}{h}, f(u(t))\right] \\
&+ \quad \limsup_{h \to 0}[d[u(t) + hf(u(t)), u_0 + hf(u_0)] \\
&- \quad d[u, u_0] + d[f(u_0), \hat{0}] \\
&\leq \quad -\beta d[u(t), u_0] + d[f(u_0), \hat{0}] \\
&= \quad -\beta m(t) + d[f(u_0), \hat{0}].
\end{aligned}$$

By condition (ii), this implies $d[f(u(t)), \hat{0}] \leq M, t \in [0, \eta)$.

This shows that for $s, t \in [0, \eta), s < t$,

$$\begin{aligned}
d[u(t), u(s)] \quad &\leq \quad d\left[u_0 + \int_0^t u(\xi)d\xi, u_0 + \int_0^s u(\xi)d\xi\right] \\
&\leq \quad d\left[\int_0^t u(\xi)d\xi, \int_0^s u(\xi)d\xi\right] \\
&\leq \quad d\left[\int_0^s u(\xi)d\xi + \int_s^t u(\xi)d\xi, \int_0^s u(\xi)d\xi\right] \\
&\leq \quad \int_s^t d[f(u(\xi)), \hat{0}]d\xi \\
&\leq \quad M(t - s), t, s \in [0, \eta).
\end{aligned}$$

It is clear from this estimate that $\lim_{t \to \eta^-} u(t)$ exists which contradicts the noncontinuability of the solution. Hence $u(t)$ exists on $[0, \infty)$. Uniqueness of solutions of (5.6.1) follows from assumption (i) because, if we assume two solutions $u(t), v(t)$ of (5.6.1) with $u(0) = v(0) = u_0$, then we get for $t \geq 0$

$$d[u(t), v(t)] \leq d[u(0), v(0)]e^{-\beta t} \equiv 0.$$

This completes the proof of Theorem 5.6.1.

Consider the fuzzy operator $S \in C[E^n, E^n]$. We are interested in finding a fuzzy fixed point of S. We define $S(u) = f(u) + u$, so that if (5.6.1) possesses a fuzzy constant solution, then that solution will be the desired fixed point. We have the following result to that effect.

Theorem 5.6.2. *Let the assumptions of Theorem 5.6.1 hold with $S(u) = f(u) + u$. Then there exists a $u^* \in E^n$ such that $Su^* = u^*$.*

Proof. By Theorem 5.6.1, there exists a unique solution $u(t, u_0)$ existing on $[0, \infty)$ for every $u_0 \in E^n$. Set $T(t)u_0 = u(t, u_0)$, $t \geq 0$. Since (5.6.1) is an autonomous differential equation, $T(t)$ defines a one-parameter family of nonlinear operators which satisfy the semigroup property. Moreover in view of condition (i), if $u(t, u_0)$, $v(t, v_0)$ are the two solutions of (5.6.1), we get by Corollary 3.4.3, the estimate

$$d[u(t, u_0, v(t, v_0)] \leq d[u_0, v_0]e^{-\beta t}, \quad t \geq 0;$$

which means

$$d[T(t)u_0, T(t)v_0] \leq d[u_0, v_0]e^{-\beta t}, \quad t \geq 0.$$

Choose $t^* > 0$ such that $e^{-\beta t^*} \leq \frac{1}{2}$. Then

$$d[T(t^*)u_0, T(t^*)v_0] \leq \frac{1}{2}d[u_0, v_0].$$

The contraction mapping theorem then shows that there exists a unique fixed point u^* of $T(t^*)$, that is $T(t^*)u^* = u^*$. We shall show that u^* is a fixed point of $T(t)$ for each $t \geq 0$. Since $T(t), T(t^*)$ commute because of the semigroup property, we then get

$$
\begin{aligned}
d[T(t)u^*, u^*] &= d[T(t)T(t^*)u^*, T(t^*)u^*] \\
&= d[T(t^*)T(t)u^*, T(t^*)u^*] \\
&\leq \frac{1}{2}d[T(t)u^*, u^*].
\end{aligned}
$$

This contradiction proves that $T(t)u^* = u^*$ for $t \geq 0$ which implies u^* is a solution of (5.6.1), that is, $0 = f(u^*)$. Since $Su = u + f(u)$, it follows that $Su^* = u^*$ and the proof is complete.

5.7   Boundary Value Problem

We shall discuss, in this section, a couple of results relative to boundary value problems of fuzzy differential equations. Consider the boundary value problem

$$u'' = f(t, u, u'), \tag{5.7.1}$$

$$u(t_1) = u_1, \quad u(t_2) = u_2, \quad t_1, t_2 \in J = [a, b], \tag{5.7.2}$$

where $f \in C[J \times E^n \times E^n, E^n]$.

We observe that there exist a unique solution of

$$u'' = h(t), \tag{5.7.3}$$

$$u(t_1) = 0, \quad u(t_2) = 0 \tag{5.7.4}$$

where $h \in C[J, E^n]$, which can be written in the form

$$u(t) = \int_{t_1}^{t_2} G(t, s)h(s)ds, \tag{5.7.5}$$

where Green's function G is given by

$$G(t, s) = \begin{cases} \frac{(t_2-t)(s-t_1)}{(t_2-t_1)}, & t_1 \le s \le t \le t_2, \\ \frac{(t_2-s)(t-t_1)}{(t_2-t_1)}, & t_1 \le t \le s \le t_2. \end{cases} \tag{5.7.6}$$

Hence the solution of (5.7.1) and (5.7.2) takes the form

$$u(t) = \int_{t_1}^{t_2} G(t, s)f(s, u(s), u'(s))ds + w(t),$$

where $w(t)$ satisfies $w'' = 0, w(t_1) = u_1, w(t_2) = u_2$.

Let us recall some properties of $G(t, s)$, namely,

$$\int_{t_1}^{t_2} |G(t, s)|ds \le \frac{(t_2 - t_1)^2}{8},$$

and

$$\int_{t_1}^{t_2} |G_t(t, s)|ds \le \frac{(t_2 - t_1)}{2}.$$

We are now in a position to prove an existence and uniqueness result for (5.7.1) and (5.7.2) using the contraction mapping principle.

Theorem 5.7.1. *Let $f \in C[J \times E^n \times E^n, E^n]$ and*

$$d[f(t, u, u'), f(t, v, v')] \leq Kd[u, v] + Ld[u', v'],$$

for $(t, u, u'), (t, v, v') \in J \times E^n \times E^n$, where $K, L > 0$. Then

$$\frac{K(t_2 - t_1)^2}{8} + \frac{L(t_2 - t_1)}{2} < 1,$$

implies that the boundary value problem (5.7.1) and (5.7.2) has a unique solution.

Proof.   Consider the space $\mathcal{C} = C^1[[t_1, t_2], E^n]$ with the metric

$$H(u, v) = \max_{t_1 \leq t \leq t_2} \left[Kd[u(t), v(t)] + Ld[u'(t), v'(t)] \right].$$

Define the operator $T : \mathcal{C} \to \mathcal{C}$ by

$$Tu(t) = \int_{t_1}^{t_2} G(t, s)f(s, u(s), u'(s))ds + w(t).$$

Then using the properties of the metric d, we get successively

$$
\begin{aligned}
d[Tu(t), Tv(t)] &\leq \int_{t_1}^{t_2} |G(t, s)|d[f(s, u(s), u'(s)), f(s, v(s), v'(s))]ds \\
&\leq H[u, v] \int_{t_1}^{t_2} |G(t, s)|ds \\
&\leq \frac{(t_2 - t_1)^2}{8} H[u, v],
\end{aligned}
$$

and

$$
\begin{aligned}
d[(Tu)'(t), (Tv)'(t)] &\leq \int_{t_1}^{t_2} |G_t(t, s)|d[f(s, u(s), u'(s)), f(s, v(s), v'(s))]ds \\
&\leq H[u, v] \int_{t_1}^{t_2} |G_t(t, s)|ds \\
&\leq \frac{(t_2 - t_1)}{2} H[u, v].
\end{aligned}
$$

Consequently, we have

$$H[Tu, Tv] \leq \left[\frac{(t_2 - t_1)^2}{8} + \frac{(t_2 - t_1)}{2} \right] H[u, v].$$

In view of the assumption, we get

$$H[Tu, Tv] \leq \beta H[u, v],$$

where $\beta = \frac{K(t_2-t_1)^2}{8} + \frac{L(t_2-t_1)}{2} < 1$. Hence the contraction mapping principle shows there exists a unique fixed point u of T and therefore $u = u(t)$ is the solution of the boundary value problem (5.4.1) and (5.7.2). The proof is complete.

Let us next consider the linear nonhomogeneous system of fuzzy differential equations

$$u' = A(t)u + f(t), \tag{5.7.7}$$

where $A(t)$ is an $N \times N$ continuous matrix of real functions on J, and $f \in C[J, E^{nN}]$, $U = (u_1, u_2, ..., u_N)$ such that $u_i \in E^n$ for each $1 \leq i \leq N$ so that $U \in E^{nN}$. Let $T : C \to E^{nN}$ be a continuous linear operator where $C = C[J_+, E^{nN}]$. Consider the boundary condition

$$TU(t) = r, \quad r \in E^{nN}. \tag{5.7.8}$$

Then we can prove the following result using the variation of parameters formula (see Section 4.8).

Theorem 5.7.2. *The boundary value problem (5.7.7) and (5.7.8) has a unique solution $U(t) \in E^{nN}$ for every $r \in E^{nN}$ and $f \in C[J_+, E^{nN}]$ if and only if the corresponding homogeneous linear boundary value problem*

$$U' = A(t)U, \quad TU(t) = \hat{0}, \tag{5.7.9}$$

has only the trivial solution $U(t) = \hat{0}$.

Proof.   Let $Y(t)$ be the fundamental solution of

$$U' = A(t)U, \quad U(0) = I \quad \text{(unit matrix)}.$$

The general solution (5.7.7) with $U(0) = c \in E^{nN}$ is given by

$$U(t) = Y(t)c + U_0(t)$$

where

$$U_0(t) = Y(t) \int_0^t Y^{-1}(s)f(s)ds, \quad t \in J,$$

is a solution of (5.7.7) with $U(0) = \hat{0}$ and $c \in E^{nN}$ arbitrary. Then the boundary condition (5.7.8) takes the form

$$TU(t)c = U_0(t) = r$$

or

$$[T^k U(t)]c = r - TU_0(t). \qquad (5.7.10)$$

Then (5.7.10) has a unique solution for any $r \in E^{nN}$ if and only if $\det[T^k U(t)] \neq 0$, that is if and only if (5.7.9) has only the trivial solution. When (5.7.10) is satisfied the solution of (5.7.7) and (5.7.8) is uniquely represented by the formula

$$U(t) = Y(t)[T^k U(t)]^{-1} r - TU_0(t) + U_0(t).$$

The proof is complete.

5.8 Fuzzy Equations of Volterra Type

We shall, in this section, consider first the following fuzzy integral equation of Volterra type

$$u(t) = f(t) + \int_{t_0}^{t} K(t, s, u(s)) ds, \qquad (5.8.1)$$

where $K \in C[J \times J \times E^n, E^n]$, $f \in C[J, E^n]$ and $J = [t_0, t_0 + a]$, $t_0 \geq 0$, $a > 0$. We shall be content in proving an existence and uniqueness theorem concerning (5.8.1) via the contraction mapping principle.

Theorem 5.8.1. *Assume that $f \in C[J, E^n]$, $K \in C[J \times J \times E^n, E^n]$ and for (t, s, u), $(t, s, v) \in J \times J \times E^n$,*

$$d[K(t, s, u), K(t, s, v)] \leq Ld[u, v], \quad L > 0. \qquad (5.8.2)$$

Then there exists a unique solution $u(t)$ on J for (5.8.1).

Proof. Let $C[J, E^n]$ denote the set of all continuous functions from J to E^n. Define the weighted metric

$$H[u, v] = \sup_J d[u(t), v(t)] e^{-\lambda t},$$

for $u, v \in C[J, E^n]$, $\lambda > 0$ to be chosen later. Since (E^n, d) is a complete metric space, the usual argument shows that $(C[J, E^n], H)$ is also a complete metric space. Define the mapping T by the relation

$$(Tu)(t) = f(t) + \int_{t_0}^{t} K(t, s, u(s)) ds, \quad t \in J.$$

Then by Corollary 2.4.2, $Tu \in C[J, E^n]$. Moreover, assumption (5.8.2) yields, using the properties of the metric d,

$$d[(Tu)(t), (Tv)(t)] \leq \int_{t_0}^t d[K(t, s, u(s)), K(t, s, v(s))]ds$$

$$\leq \int_{t_0}^t Ld[u(s), v(s)]ds, \quad t \in J.$$

This, in turn, implies that

$$e^{-\lambda t}d[(Tu)(t), (Tv)(t)] \leq Le^{-\lambda t}H[u, v]\int_{t_0}^t e^{\lambda s}ds$$

$$\leq \frac{L}{\lambda}H[u, v].$$

Thus choosing $\lambda = 2L$, we get

$$H[Tu, Tv] \leq \frac{1}{2}H[u, v].$$

The contraction mapping principle then assures the existence of a unique fixed point of T, say $u^* \in C[J, E^n]$, which proves that $u^*(t)$ is the unique solution of the fuzzy integral equation (5.8.1) on J. The proof is complete.

Let us next consider the abstract Volterra operator which includes several types of functional equations such as ordinary fuzzy differential equations, fuzzy delay differential equations and integral and integro-differential equations of fuzzy type. Such operators are known as causal or nonanticipative operators.

Let $E = C[[t_0, T], E^n]$ and $V : E \to E$. We say that V is a causal operator if $u, v \in E$ such that $u(s) = v(s)$ for $t_0 \leq s \leq t$, then $(Vu)(s) = (Vv)(s)$ for $t_0 \leq s \leq t$, with $t < T$. We consider the fuzzy Volterra functional equation

$$u'(t) = (Vu)(t), \quad u(t_0) = u_0 \in E^n, \quad t \in [t_0, T], \qquad (5.8.3)$$

where V satisfies the generalized Lipschitz-type condition

$$\int_{t_0}^t d[(Vu)(s), (Vv)(s)]ds \leq \int_{t_0}^t \lambda(s)d[u(s), v(s)]ds, \qquad (5.8.4)$$

where $\lambda \in C[[t_0, T], R_+], u, v \in E$. By the contraction mapping principle, we shall show that

$$u(t) \to u_0 + \int_{t_0}^t (Vu)(s)ds \qquad (5.8.5)$$

and $H_g(u,v) = \sup_{t_0 \le t \le T} g(t) d[u(t), v(t)]$ where g is continuous from $[t_0, T]$ into $(0, \infty)$. If we denote the right-hand side of (5.8.5) by $(Qu)(t)$, it is clear that Q takes values in E and hence we obtain

$$d[(Qu)(t), (Qv)(t)] \le \int_{t_0}^{t} \lambda(s) d[u(s), v(s)] ds. \tag{5.8.6}$$

In order to use the weighted metric in Eg, we shall proceed as follows. Let $g(t) = e^{\alpha \int_{t_0}^{t} \lambda(s) ds, \alpha > 1}$. Then

$$
\begin{aligned}
d[(Qu)(t), (Qv)(t)] &\le \int_{t_0}^{t} \lambda(s) e^{\alpha \int_{t_0}^{s} \lambda(\xi) d\xi} d[u(s), v(s)] e^{-\alpha \int_{t_0}^{s} \lambda(\xi) d\xi} ds \\
&\le \sup_{t_0 \le t \le T} \left(d[u(t), v(t)] e^{-\alpha \int_{t_0}^{t} \lambda(\xi) d\xi} \right) \int_{t_0}^{t} \lambda(s) e^{\alpha \int_{t_0}^{s} \lambda(\xi) d\xi} ds \\
&\le \frac{1}{\alpha} e^{\alpha \int_{t_0}^{t} \lambda(\xi) d\xi} H_g[u, v].
\end{aligned}
$$

From the foregoing estimate, we get

$$
\begin{aligned}
H_g[Qu, Qv] &= d[(Qu)(t), (Qv)(t)] e^{-\alpha \int_{t_0}^{t} \lambda(\xi) d\xi} \\
&\le \frac{1}{\alpha} H_g[u, v]
\end{aligned}
$$

which shows Q is a contraction. We need to show that $QEg \subset Eg$. For this purpose, we need the assumption

$$d[(V\hat{0})t, \hat{0}] \le K\lambda(t) \exp\left[\alpha \int_{t_0}^{t} \lambda(s) ds \right]. \tag{5.8.7}$$

Then we obtain

$$
\begin{aligned}
d[(Qu)t, \hat{0}] &\le d[u_0, \hat{0}] + \int_{t_0}^{t} d[(Vu)(s), (V\hat{0})(s)] ds + \int_{t_0}^{t} d[(V\hat{0})s, \hat{0}] ds \\
&\le d[u_0, \hat{0}] + \int_{t_0}^{t} \lambda(s) d[u(s), \hat{0}] ds + K \int_{t_0}^{t} \lambda(s) e^{\alpha \int_{t_0}^{s} \lambda(\xi) d\xi} ds.
\end{aligned}
$$

Since $u \in Eg$, it is easy to get from the preceding estimate

$$d[(Qu)(t), \hat{0}] \le d[u_0, \hat{0}] + \frac{1}{\alpha}(A_u + K) \exp\left(\alpha \int_{t_0}^{t} \lambda(\xi) d\xi \right),$$

showing that $QEg \subset Eg$. Hence the operator Q takes Eg into itself and it is a contraction mapping and as a result, there exists a unique solution $u(t)$ on $[t_0, T]$ of (5.8.3). This completes the proof of the following result.

Theorem 5.8.2. *Consider (5.8.3) and assume that (5.8.4) and (5.8.7) hold. Then there exists a unique solution $u(t)$ on $[t_0, T]$ for (5.8.3).*

5.9   A New Concept of Stability

Following the notation of Section 4.6, we consider the fuzzy differential system

$$u' = f(t, u), \quad u(t_0) = u_0, \quad t \geq t_0, \tag{5.9.1}$$

where $f \in C[\mathcal{R}_+ \times E^{nN}, E^{nN}]$ and $u_0 \in E^{nN}$. We employ the metric space (E^{nN}, d_0) where $d_0[u, v] = \sum_{i=1}^{N} d[u_i, v_i]$, $u_i, v_i \in E^n$ for each i, $1 \leq i \leq N$. We also utilize the generalized metric space $(E^{nN} D)$ where

$$D[u, v] = (d[u_1, v_1], d[u_2, v_2], \ldots, d[u_N, v_N]).$$

We need the following known results. See Lakshmikantham and Leela [61].

Hereafter, the inequalities between vectors in $\mathcal{R}^d$ are to be understood componentwise.

Theorem 5.9.1. *Let $g \in C[\mathcal{R}_+ \times \mathcal{R}_+^d \times \mathcal{R}_+^d, \mathcal{R}^d]$, $g(t, w, \xi)$ be quasi-monotone nondecreasing in w for each (t, ξ) and monotone nondecreasing in ξ for each (t, w). Suppose further that $r(t) = r(t, t_0, w_0)$ is the maximal solution of*

$$w' = g(t, w, w), \quad w(t_0) = w_0 \geq 0, \tag{5.9.2}$$

existing on $[t_0, \infty)$. Then the maximal solution $\mathcal{R}(t) = \mathcal{R}(t, t_0, w_0)$ of

$$w' = g(t, w, r(t)), \quad w(t_0) = w_0 \geq 0, \tag{5.9.3}$$

exists on $[t_0, \infty)$ and

$$r(t) \equiv \mathcal{R}(t), \quad t \geq t_0. \tag{5.9.4}$$

Theorem 5.9.2. *Assume that the function $g(t, w, \xi)$ satisfies the conditions of Theorem 5.9.1. Then $m \in C[\mathcal{R}_+, \mathcal{R}_+^d]$ and*

$$D^+ m(t) \leq g(t, m(t), \xi), \quad t \geq t_0. \tag{5.9.5}$$

Then for all $\xi \leq r(t)$, it follows that

$$m(t) \leq r(t), \quad t \geq t_0.$$

We can now state the needed comparison results in terms of suitable Lyapunov-like functions which can easily be proved. For this purpose, we let

$$\Omega = [\sigma \in C^1[\mathcal{R}_+, \mathcal{R}_+] : \sigma(t_0) = t_0 \text{ and } w(t, \sigma, \sigma') \leq r(t), t \geq t_0], \tag{5.9.6}$$

where $w \in C[\mathcal{R}_+^2 \times \mathcal{R}, \mathcal{R}_+^d]$ and $r(t)$ is the maximal solution of (5.9.2).

Theorem 5.9.3. *Assume that for some $\sigma \in \Omega$, there exists a V such that $V \in C[\mathcal{R}_+^2 \times E^{nN} \times E^{nN}, \mathcal{R}_+^d]$,*

$$|V(t, \sigma, u_1, v_1) - V(t, \sigma, u_2, v_2)|$$

$$\leq A[D[u_1, u_2] + D[v_1, v_2]],$$

where A is an $N \times N$ positive matrix, and

$$D^+V(t, \sigma, u, v)$$
$$= \limsup_{h \to 0^+} \frac{1}{h}[V(t+h, \sigma(t+h), u + hf(t, u), v + hf(\sigma, v)\sigma')$$
$$-V(t, \sigma, u, v)]$$
$$\leq g(t, V(t, \sigma, u, v), w(t, \sigma, \sigma')),$$

where $g(t, w, \xi)$ satisfies the conditions of Theorem 5.9.1.
Then $V(t_0, \sigma(t_0), u_0, v_0) \leq w_0$ implies that

$$V(t, \sigma(t), u(t, t_0, u_0), v(\sigma(t), t_0, v_0)) \leq r(t, t_0, w_0), \quad t \geq t_0.$$

Let us now introduce the new concept of stability. Let $v(t, t_0, v_0)$ be the given unperturbed solution of (5.9.1) on $[t_0, \infty)$, and let $u(t, t_0, u_0)$ be any perturbed solution of (5.9.1) on $[t_0, \infty)$. Then Lyapunov stability (LS) compares the phase space positions of the unperturbed and perturbed solutions at exactly simultaneous instants; namely

$$d_0[ut, t_0, u_0), v(t, t_0, v_0)] < \epsilon, \quad t \geq t_0, \tag{LS}$$

which is too restrictive a requirement from the physical point of view. Orbital stability (OS), on the other hand, compares phase space positions of the same solutions at any two unrelated times; namely,

$$\inf_{s \in [t_0, \infty)} d_0[u(t, t_0, u_0), v(s, t_0, v_0)] < \epsilon, \quad t \geq t_0. \tag{OS}$$

In this case, the measurement of time is completely irregular and therefore (OS) is too loose a demand.

We therefore need a new notion unifying (LS) and (OS) which would lead to concepts between them that could be physically significant. This is precisely what we plan to do below.

Let E denote the space of all functions from $\mathcal{R}_+ \to \mathcal{R}_+$, each function $\sigma(t) \in E$ representing a clock. Let us call $\sigma(t) = t$ the perfect clock. Let $\tau-$ be any topology in E. Given the solution $v(t, t_0, v_0)$ of (5.9.1) existing on $[t_0, \infty)$, we define following Messera [73] the new concept of stability as follows.

Definition 5.9.1. *The solution $u(t, t_0, v_0)$ of (5.9.1) is said to be*

(1) *τ-stable, if, given $\epsilon > 0$, $t_0 \in \mathcal{R}_+$, there exist a $\delta = \delta(t_0, \epsilon) > 0$ and a τ-neighborhood of N of the perfect clock satisfying the condition that $d_0[u_0, v_0] < \delta$ implies $d_0[u(t, t_0, u_0), v(\sigma(t), t_0, v_0)] < \epsilon$, $t \geq t_0$ where $\sigma \in N$;*

(2) *τ-uniformly stable, if δ in (1) is independent of t_0.*

(3) *τ-asymptotically stable, if (1) holds and given $\epsilon > 0$, $t_0 \in \mathcal{R}_+$, there exist a $\delta_0 = \delta_0(t_0) > 0$, a τ-neighborhood N of the perfect clock, and a $T = T(t_0, \epsilon) > 0$ such that*

$$d_0[u_0, v_0] < \delta_0 \text{ implies } d_0[u(t, t_0, u_0), v(\sigma(t), t_0, v_0)] < \epsilon,$$

$$t \geq t_0 + T,$$

where $\sigma \in N$;

(4) *τ-uniformly asymptotically stable, if δ_0 and T are independent of t_0.*

We note that a partial ordering of topologies induces a corresponding partial ordering of stability concepts.

Let us consider the following topologies of E:

(τ_1) the discrete topology, where every set of E is open;

(τ_2) the chaotic topology, where the open sets are only the empty set and the entire clock space E;

(τ_3) the topology generated by the base

$$U_{\sigma_0, \epsilon} = \left[\sigma \in E : \sup_{t \in [t_0, \infty)} |\sigma(t) - \sigma_0(t)| < \epsilon \right];$$

(τ_4) the topology defined by the base

$$u_{\sigma_0, \epsilon} = \left[\sigma \in C^1[\mathcal{R}_+, \mathcal{R}_+] : |\sigma(t_0) - \sigma_0(t_0)| < \epsilon \text{ and} \right.$$

$$\left. \sup_{t \in [t_0, \infty)} |\sigma'(t) - \sigma_0'(t)| < \epsilon \right].$$

It is easy to see that the topologies τ_3, τ_4 lie between τ_1 and τ_2. Also, an obvious conclusion is that if the unperturbed motion $v(t, t_0, v_0)$ is the trivial solution, then (OS) implies (LS).

In τ_1-topology, one can use the neighborhood consisting solely of the perfect clock $\sigma(t) = t$ and therefore Lyapunov stability follows immediately from the existing results.

Define $B = B[t_0, v_0] = v([t_0, \infty), t_0, v_0)$ and suppose that B is closed. Then the stability of the set B can be considered in the usual way in terms of Lyapunov functions since

$$\rho[u(t, t_0, u_0), B] = \inf_{s \in [t_0, \infty)]} d_0[u(t, t_0, u_0), v(s, t_0, v_0)];$$

denoting the infimum for each t by s_t and defining $\sigma(t) = s_t$ for $t > t_0$, we see that $\sigma \in E$ in the τ_2-topology. We therefore obtain orbital stability of the given solution $v(t, t_0, v_0)$ in terms of the τ_2-topology.

To investigate the results corresponding to the τ_3 and τ_4 topologies, we shall utilize the comparison Theorem 5.9.3 and suitably modify the proofs of standard stability results.

Theorem 5.9.4. *Let condition (i) of Theorem 5.9.3 be satisfied. Suppose further that*

(a) $b(d_0[u, v]) \leq \sum_{i=1}^{d} v_i(t, \sigma, u, v) \leq a(t, \sigma, d_0[u, v])$,

(b) $d(|t - \sigma(t)|) \leq \sum_{i=1}^{d} w_i(t, \sigma, \sigma')$,

where $a(t, \sigma, \cdot)$, $b(\cdot)$ and $d(\cdot) \in \mathcal{K} = [a \in C[\mathcal{R}_+, \mathcal{R}_+], a(0) = 0$ and $a(\eta)$ is increasing in η.

Then the stability properties of the trivial solution of (5.9.2) imply the corresponding τ_3-stability properties of the fuzzy differential system (5.9.1) relative to the given solution $v(t, t_0, v_0)$.

Proof. Let $v(t) = v(t, t_0, v_0)$ be the given solution of (5.9.1) and let $0 < \epsilon$ and $t_0 \in \mathcal{R}_+$ be given. Suppose that the trivial solution of (5.9.2) is stable. Then given $b(\epsilon) > 0$ and $t_0 \in \mathcal{R}_+$, there exists a $\delta_1 = \delta_1(t_0, \epsilon) > 0$ such that

$$0 \leq \sum_{i=1}^{d} w_{i0} < \delta_1 \quad \text{implies} \quad \sum_{i=1}^{d} w_i(t, t_0, w_0) < b(\epsilon), \ t \geq t_0,$$

where $w(t, t_0, w_0)$ is any solution of (5.9.2). We set $w_0 = V(t_0, \sigma(t_0), u_0, v_0)$ and choose $\delta = \delta(t_0, \epsilon)$ $\eta = \eta(\epsilon)$ satisfying

$$a(t_0, \sigma(t_0), \delta) < \delta_1 \quad \text{and} \quad \eta = d^{-1}(b(\epsilon)). \tag{5.9.7}$$

Using (b) and the fact that $\sigma \in \Omega$, we have

$$
\begin{aligned}
d(|t - \sigma|) &\leq \sum_{i=1}^{d} w_i(t, \sigma, \sigma') \\
&\leq \sum_{i=1}^{d} r_i(t, t_0, w_0) \\
&\leq \sum_{i=1}^{d} r_i(t, t_0, \delta_1) < b(\epsilon).
\end{aligned}
$$

It then follows that $|t - \sigma(t)| < \eta$ and hence $\sigma \in N$. We claim that whenever

$$
d_0[u_0, v_0] < \delta \quad \text{and} \quad \sigma \in N,
$$

it follows that

$$
d_0[u(t, t_0, u_0), v(\sigma(t), t_0, v_0] < \epsilon, \quad t \geq t_0.
$$

If this were not true, there would exist a solution $u(t, t_0, u_0)$ and a $t_1 > t_0$ such that

$$
\begin{aligned}
d_0[u(t_1, t_0, u_0), v(\sigma(t_1), t_0, v_0)] &= \epsilon \quad \text{and} \\
d_0[u(t, t_0, u_0), v(\sigma(t), t_0, v_0)] &\leq \epsilon
\end{aligned}
\tag{5.9.8}
$$

for $t_0 \leq t \leq t_1$. Then by Theorem 5.9.3, we get for $t_0 \leq t \leq t_1$,

$$
V(t, \sigma(t), u(t, t_0, u_0), v(\sigma(t), t_0, v_0)) \leq r(t, t_0, V(t_0, \sigma(t_0), u_0, v_0))),
$$

where $r(t, t_0, w_0)$ is the maximal solution of (5.9.2). It then follows from (5.9.2), (5.9.8), using (a), that

$$
\begin{aligned}
b(\epsilon) &= b(d_0[u(t_1), v(\sigma(t_1))]) \\
&\leq \sum_{i=1}^{d} V_i(t_1, \sigma(t_1), u(t_1), v(\sigma(t_1))) \\
&\leq \sum_{i=1}^{d} r_i(t_1, t_0, V(t_0, \sigma(t_0), u_0, v_0)) \\
&\leq \sum_{i=1}^{d} r_i(t_1, t_0, a(t_0, \sigma(t_0), \delta_1)) < b(\epsilon),
\end{aligned}
$$

a contradiction, which proves τ_3-stability.

Suppose next that the trivial solution of (5.9.2) is asymptotically stable. Then it is stable and given $b(\epsilon) > 0$, $t_0 \in \mathcal{R}_+$, there exist $\delta_{01} = \delta_{01}(t_0) > 0$ and $T = T(t_0, \epsilon) > 0$ satisfying

$$0 \le \sum_{i=1}^{d} w_{0i} < \delta_{10} \text{ implies } \sum_{i=1}^{d} w_i(t, t_0, w_0) < b(\epsilon), \; t \ge t_0 + T. \qquad (5.9.9)$$

The τ_3-stability yields, taking $\epsilon = \rho > 0$ and designating $\delta_0(t_0) = \delta(t_0, \rho)$,

$$d_0[u_0, v_0] < \delta_0 \text{ implies } d_0[u(t), v(\sigma(t))] < \rho, \; t \ge t_0$$

for every σ such that $|t - \sigma| < \eta(\rho)$. This means that by Theorem 5.9.3

$$V(t, \sigma(t), u(t), v(t)) \le r(t, t_0, \delta_{10}), \quad t \ge t_0. \qquad (5.9.10)$$

In view of (5.9.9), we find that

$$\sum_{i=1}^{d} r_i(t, t_0, \delta_{10}) < b(\epsilon), \quad t \ge t_0 + T,$$

which in turn implies

$$\begin{aligned} d[|t - \sigma(t)|] &\le \sum_{i=1}^{d} w_i(t, \sigma, \sigma') \\ &\le \sum_{i=1}^{d} r_i(t, t_0, \delta_{10}) \\ &< b(\epsilon), \quad t \ge t_0 + T. \end{aligned}$$

Thus $|t - \sigma(t)| < d^{-1}b(\epsilon) = \eta(\epsilon)$, $t \ge t_0 + T$. Hence there exists a $\sigma \in N$ satisfying

$$\begin{aligned} d_0[u(t), v(\sigma(t))] &\le \sum_{i=1}^{d} V_i(t, \sigma(t), u(t), v(\sigma(t))) \\ &\le \sum_{i=1}^{d} r_i(t, t_0, \delta_{10}) < b(\epsilon), \quad t \ge t_0 + T, \end{aligned}$$

which yields

$$d_0[u(t), v(\sigma(t))] < \epsilon, \quad t \ge t_0 + T,$$

whenever $d_0[u_0, v_0] < \delta_0$ and $\sigma \in N$. This proves τ_3-asymptotic stability and the proof is complete.

To obtain sufficient conditions for τ_4-stability, we need to replace (b) in Theorem 5.9.4 by

(c) $d[|1 - \sigma'(t)|] \le \sum_{i=1}^{d} w_i(t, \sigma, \sigma')$,

and then mimic the proof with suitable modifications. We omit the details to avoid monotony.

It would be interesting to obtain different sets of sufficient conditions as well as discover other topologies that would be of interest.

5.10   Notes and Comments

Section 5.2 introduces fuzzy difference equations and the results presented are new, and are based on the corresponding results in difference equations. See Lakshmikantham and Trigiante [69]. The contents of Section 5.3 dealing with impulsive fuzzy differential equations are adapted from Lakshmikantham and McRae [65]. Refer to Lakshmikantham, Bainov and Simeonov [58] for details for impulsive differential equations.

The results of Section 5.4 that initiate the theory of fuzzy differential equations with delay are new. Concerning the new results described in Section 5.5 relative to fuzzy hybrid systems, see Lakshmikantham and Liu [63] and Rizzo [101] on which the present results are modelled.

For the contents of Section 5.6 that develop the fixed points of fuzzy mappings using the theory of fuzzy differential equations refer to Lakshmikantham and Vatsala [70]. The boundary value problems of fuzzy differential equations introduced in Section 5.7 are adapted from Lakshmikantham, Murty and Turner [68]. The typical results presented in Section 5.8 concerning the fuzzy operator equations of Volterra type are new and are based on Corduneanu [13]. Section 5.9 consists of results taken from Lakshmikantham [57]. For related results see Seikkala [105], Corduneanu [13, 12], and Ma et al. [72].

Chapter 6

Fuzzy Differential Inclusions

6.1   Introduction

Recall that the theory of fuzzy differential equations (FDEs) considered so far utilizes the Hukuhara derivative (H-derivative) for the formulation. We have investigated, in the previous chapters, several basic results of fuzzy differential equations via the comparison principle in the metric space (E^n, d) with no complete linear structure. This approach for fuzzy differential equations which employs the H-derivative suffers from a disadvantage because the solution $x(t)$ of an FDE has the property that $\mathrm{diam}[x(t)]^\beta$ is nondecreasing in time, that is, the solution is irreversible in probabilistic terms. Consequently, it has been recently realized that this formulation of FDEs cannot really reflect any rich behavior of solutions of ODEs such as stability, periodicity and bifurcation, and therefore is not well-suited for modeling.

Alternative approaches have recently been introduced by Buckley and Feuring [8], Vorobiev and Seikkala [112], and Hüllermeier [40]. A different and interesting framework suggested by Hüllermeier is more general than the others. It is based on a family of differential inclusions at each β-level, $0 \le \beta \le 1$, namely

$$x'(t) \in [G(t, x(t))]^\beta, \quad x(0) = [x_0]^\beta, \qquad (6.1.1)$$

where $[G(\cdot, \cdot)]^\beta \colon \mathcal{R} \times \mathcal{R}^n \to \mathcal{K}_c^n$, the space of nonempty compact convex subsets of $\mathcal{R}^n$. The idea is that the set of all such solutions $S_\beta(x_0, T)$ would be the β-level of a fuzzy set $S(x_0, T)$, in the sense that all attainable sets $\mathcal{A}(x_0, t)$, $0 < t \le T$, are levels of the fuzzy set $\mathcal{A}_\beta(x_0, t)$ on $\mathcal{R}^n$. This framework captures both vagueness (uncertainty) and the rich properties of differential inclusions in one and the same technique. For example, with this interpretation, the FDE

$$x'(t) = tx(t)u + v,$$

where u, v are constant fuzzy numbers, is the family of inclusions

$$x'(t) \in tx(t)[u]^\beta + [v]^\beta, \quad 0 \le \beta \le 1, \tag{6.1.2}$$

where $[u]^\beta$, $[v]^\beta$ are β-level sets. In this chapter, we shall use the notation for β-level sets, the usual $[u]^\beta$ and u_β as convenient. Let us provide a simple one-dimensional example to illustrate the situation. Let $(c; d)_S$ denote the symmetric triangular fuzzy number with the interval $[c, d]$ as its support and let $x(t)$ be a fuzzy number valued function of time. Consider the FDE initial value problem

$$x'(t) = -2x, \quad x(0) = X^0 = (0; 1)_S. \tag{6.1.3}$$

Write the β-level set of $x(t)$ as the compact interval $x_\beta(t) = [x_\beta^L(t), x_\beta^R(t)]$ and note that $-2x_\beta = [-2x_\beta^R, -2x_\beta^L]$, while $x_\beta(0) = [\beta/2, 1 - \beta/2]$. Writing $\xi_\beta(t)$ as the vector with components $x_\beta^L(t)$, $x_\beta^R(t)$, obtain the ordinary initial value problem

$$\xi_\beta'(t) = \begin{pmatrix} 0 & -2 \\ -2 & 0 \end{pmatrix} \xi_\beta, \quad \xi_\beta(0) = \begin{pmatrix} \beta/2 \\ 1 - \beta/2 \end{pmatrix},$$

for $0 \le \beta \le 1$. It is easy to see that

$$\xi_\beta(t) = \frac{1}{2}e^{-2t} \begin{pmatrix} 1 \\ 1 \end{pmatrix} + \frac{\beta - 2}{2}e^{2t} \begin{pmatrix} 1 \\ -1 \end{pmatrix},$$

that is,

$$x_\beta(t) = [e^{-2t}/2 + (\beta - 1)e^{2t}/2, e^{-2t}/2 + (1 - \beta)e^{2t}/2].$$

Hence, the solution to (6.1.3) is $x(t) = e^{-2t}/2 + (-e^{2t}/2; e^{2t}/2)_S$. Thus, (6.1.3) has an unstable solution in contrast to the behavior of the associated crisp problem $z'(t) = -2z(t)$, $z(0) = 1/2$, which has solution $z(t) = e^{-2t}/2$. So giving the initial condition some uncertainty by fuzzification has totally changed the qualitative behavior of the solution. Indeed, an arbitrarily small fuzzification $x(0) = (1/2 - \epsilon; 1/2 + \epsilon)_S$ has the same effect, the solution being $x(t) = e^{-2t}/2 + (-\epsilon e^{2t}; \epsilon e^{2t})_S$, although, as $\epsilon \to 0+$, the crisp solution is the limit.

If, on the other hand, we take the β-level set of $x(0)$ as the interval $[x_{10}^\beta, x_{20}^\beta]$, we get

$$\xi_\beta(t) = \frac{1}{2}e^{-2t}\left(\frac{2x_{10}^\beta + x_{20}^\beta}{2x^{\beta^2}10 + x_{20}^\beta}\right) + \frac{1}{2}e^{2t}\left(\frac{2x_{10}^\beta - x_{20}^\beta}{x_{20}^{\beta^2} - 2x_{10}^\beta}\right).$$

If we have therefore $X_\beta^0 = [x_{10}^\beta, 2x_{10}^\beta]$, then we can see that

$$x_\beta(t) = [e^{-2t}x_{10}^\beta, e^{-2t}x_{10}^\beta],$$

which shows that the behavior of solutions matches the associated crisp problem.

The framework suggested by Hüllermeier [40] has largely overcome such an undesirable property of solutions. If the FDE (6.1.3) is replaced by the family

$$x_\beta' \in [-2x_\beta^R, -2x_\beta^I], x_\beta(0) \in X_\beta^0 - [\beta/2, 1 - \beta/2], 0 \le \beta \le 1, \qquad (6.1.4)$$

then it has a fuzzy solution set $S(X^0, \tau)$ and a fuzzy attainability set $\mathcal{A}(X^0, t)$ respectively defined by the β-level sets

$$S_\beta(X_\beta^0, \tau) = \{x(\cdot) : x(t) \subset [\beta c^{-2t}, (1 - \beta/2)e^{-2t}], 0 \le t \le \tau\}$$

$$\mathcal{A}_\beta(X_\beta^0, t) = [\beta e^{-2t}, (1 - \beta/2)e^{-2t}].$$

This matches the sort of desirable behavior a fuzzification of the crisp DE should have, namely it is asymptotically stable and approaches the crisp limit as the uncertainty becomes negligible.

In this chapter, we shall discuss the new formulation which is of very recent origin and is still in the nascent stage. We adapt mostly Diamond's work, some of which is unpublished. In Section 6.2 we shall formulate fuzzy differential inclusions (FDIs) following Hüllermeier's development. Section 6.3 describes the necessary results of the known theory of differential inclusions. Section 6.4 is devoted to FDIs and the results on stability, and periodicity are provided in the new framework. In Section 6.5, we shall discuss the variation of constants formula for fuzzy differential inequalities of linear type. In Section 6.6, we shall extend the approach to fuzzy integral inequalities. Section 6.7 deals with notes and comments.

6.2   Formulation of Fuzzy Differential Inclusions

In this section, we shall define fuzzy differential inclusions as a family of differential inclusions at each β-level, $0 \leq \beta \leq 1$. For this purpose, we need to first develop regularity of solution sets for differential inclusions which are quasi-concave with respect to a parameter.

For various appropriate conditions on $f : \mathcal{R}^{n+k+1} \to \mathcal{R}$, the initial value problem with parameter $p \in \mathcal{R}^k$,

$$x'(t) = f(t, x; p), \quad x(t_0) = x_0,$$

has solutions $x(t; t_0, x_0, p)$ which are continuous or differentiable with respect to x_0 and p. The situation with a differential inclusion

$$x'(t) \in F(t, x; p), \quad x(t_0) = x_0,$$

is somewhat different. If F is upper semicontinuous (usc) convex compact valued, the set of solutions on $[t_0, T]$, $S_p(x_0, [t_0, T])$ is compact, connected and usc in x_0 and p (see Deimling [17] and Aubin and Cellina [1]), but not generally convex. If F is only lower semicontinuous $S_p(x_0, [t_0, T])$ need not even be closed.

We shall extend the concept of quasi-concavity to multivalued functions in order to apply to fuzzy differential inclusions. Denote by $\mathcal{K}^n$ (resp. $\mathcal{K}_C^n$) the nonempty compact (resp. convex compact) subsets of $\mathcal{R}^n$, let $\Omega \subset \mathcal{R} \times \mathcal{R}^n$ be open and let I be a real compact interval. A mapping $F : \Omega \times I \to \mathcal{K}^n$ is said to be *regularly quasi-concave on* I if

(i) For all $(t, x) \in \Omega$ and $\alpha, \beta \in I$,

$$F(t, x; \alpha) \supseteq F(t, x; \beta) \quad \text{whenever} \quad \alpha \leq \beta. \qquad (6.2.1)$$

(ii) If $\{\beta_n\}$ is a nondecreasing sequence in I converging to β, then for all $(t, x) \in \Omega$

$$\bigcap_n F(t, x; \beta_n) = F(t, x; \beta). \qquad (6.2.2)$$

The definition is adapted from the usual definition for real-valued functions dropping the convexity of level sets.

Now consider the differential inclusion

$$x_\beta'(t) \in F(t, x_\beta(t); \beta) \quad \text{a.e. in} \quad t, \quad x_\beta(0) = x_0, \qquad (6.2.3)$$

where Ω is an open subset of $\mathcal{R}^{n+1}$ containing $(0, x_0)$, I is a compact interval and $F : \Omega \times I \to \mathcal{K}_C^n$. Throughout, it is assumed that all maps are *proper*, that is have nonempty images of points in their domain. The *boundedness assumption* is said to hold if there exist b, T, $M > 0$ such that the set $Q = [0, T] \times (x_0 + (b + MT)B) \subset \Omega$ where B is the unit ball of $\mathcal{R}^n$, and F maps $Q \times I$ into the ball of radius M. Denote the set of all solutions of (6.2.3) on $[0, \tau]$ by $S_\beta(x_0, \tau)$, the attainable set $\mathcal{A}_\beta(x_0, \tau) = \{x(\tau) : x(\cdot) \in S_\beta(x_0, \tau)\}$ and write $\mathcal{Z}_T(\mathcal{R}^n)$ $\{x \in C([0, T]; \mathcal{R}^n) : x' \in L^\infty([0, T]; \mathcal{R}^n)\}$. It is known that for every $x_1 \in x_0 + \mathrm{bint}B$, $S_\beta(x_1, T)$ exists, and is a compact subset of $\mathcal{Z}_T(\mathcal{R}^n)$, and each attainable section $\mathcal{A}_\beta(x_1, \tau)$, $0 < \tau \le T$, is a compact subset of $\mathcal{R}^n$, see Aubin and Cellina [1]. In fact, although these sets are not in general convex, they are acyclic which is stronger than simply connected. See De Blasi and Myjak [15].

Theorem 6.2.1. *Let $F : \Omega \times I \to \mathcal{K}_C^n$ be usc on Ω, regularly quasi-concave on J and suppose that the boundedness assumption holds. Then the mapping $\beta \mapsto \mathcal{A}_\beta(x_0, T)$ is a a regularly quasi-concave map from I to $\mathcal{K}^n$, and $\beta \mapsto S_\beta(x_0, T)$ is a regularly quasi-concave mapping from J to $\mathcal{Z}_T(\mathcal{R}^n)$.*

Proof.   Abbreviate $S_\beta = S_\beta(x_0, T)$. It is clear from (6.2.1) that for $\alpha \le \beta$, $S_\beta \subseteq S_\alpha$. Let β_n be a nondecreasing sequence converging to β. Then S_{β_n} is a decreasing sequence of compact sets and so $\cap_n S_{\beta_n} = \tilde{S}$ is nonempty and compact. Furthermore $\tilde{d}_H(S_{\beta_n}, \tilde{S}) \to 0$, where $\tilde{d}_H$ is the Hausdorff metric in $\mathcal{Z}_T(\mathcal{R}^n)$. To see that $S_\beta = \tilde{S}$, it suffices to show $\tilde{S} \subset S_\beta$ since $S_\beta \subset \tilde{S}$ is clear. For each n let $x_{\beta_n} \in S_{\beta_n}$. Since F is bounded by the ball of radius M, x'_{β_n} is bounded so $\{x_{\beta_n}\}$ is an equicontinuous family. By Theorem 0.3.4 of Aubin and Cellina [1], a subsequence $x_{\beta_n(i)}$ converges to some $v \in C([0, T], \mathcal{R}^n)$ and $x'_{\beta_n(i)}$ converges in a weak topology of $L^1([0, T], \mathcal{R}^n)$ to v' and thus weakly* in $L^\infty([0, T], \mathcal{R}^n)$ by Alaoglu's theorem. Observe that $v \in \tilde{S}$. That $v \in S_\beta$ is a consequence of the convergence theorem in Aubin and Cellina, Theorem 1.4.5 [1] and the use of F. Let N be an arbitrary neighborhood of the origin in $\Omega \times [0, 1] \times \mathcal{R}^n$ and choose $\epsilon > 0$ such that

$$\epsilon < \inf\{\|a - b\| : a \in \mathrm{Gr}(F), b \in (\mathrm{Gr}(F) + N)^c\}$$

which is possible because $\mathrm{Gr}(F)$, the graph of F, is compact. By usc, there exists a neighborhood U of $(t, v(t), \beta)$ such that for $(s, x, \alpha) \in U$, $F(s, x, \alpha) \subset F(t, v(t), \beta) + \epsilon B$. Choosing n sufficiently large $(t, x_{\beta_n}, \beta_n) \in U$ and so

$$F(t, x_{\beta_n}, \beta_n) \subset F(t, v(t), \beta) + \epsilon B$$

which means $(t, x_{\beta_n}(t), \beta_n, x'_{\beta_n}) \in \mathrm{Gr}(F) + N$. The Convergence Theorem implies $v'(t) \in F(t, v(t), \beta)$ a.e. and so $v \in S_\beta$ as required.

Example 6.2.1. *Consider the FIVP in* $\mathcal{E}^1$, $x' = -\lambda x$, $x(0) = X_0$, *where* X_0 *is a symmetric triangular fuzzy number with support* $[-1, 1]$. *When this is interpreted as a family of differential inclusions*

$$x'_\beta(t) \in -\lambda x_\beta(t), \ x_\beta(0) \in X_\beta := (1 - \beta)[-1, 1], \ 0 \le \beta \le 1, \qquad (6.2.4)$$

regular quasi-concavity is especially evident. Since $-\lambda x_\beta = \{-\lambda x_\beta\}$ *is a singleton set in* $\mathcal{K}_C^1$, *(6.2.4) becomes*

$$x'_\beta(t) = -\lambda x_\beta(t), \ x_\beta(0) \in X_\beta = (1 - \beta)[-1, 1],$$

which has solution set $S_\beta(X_\beta, t)$ *on* $[0, t]$ *comprising the functions*

$$x_\beta(t) = x_\beta(0)e^{-\lambda t}, x_\beta(0) \in X_\beta.$$

Consequently, $S_\beta(X_\beta, t) = (1 - \beta)e^{-\lambda t}[-1, 1]$. *Obviously,*

$$\beta \mapsto S_\beta(X_\beta, T) = \{(1 - \beta e^{-\lambda t}[-1, 1] : t \in [0, T]\}$$

is regularly quasi-concave, as is $\beta \mapsto \mathcal{A}_\beta(X_\beta, T) = (1 - \beta)e^{\lambda T}[-1, 1]$.

Let $\mathcal{D}^n$ denote the set of usc normal fuzzy sets on $\mathcal{R}^n$ with compact support. Clearly, $E^n \subset \mathcal{D}^n$, since elements of $\mathcal{D}^n$ have nonempty compact, but not necessarily convex β-level sets. The following characterization of elements of $\mathcal{D}^n$ is required.

Theorem 6.2.2. (Stacking theorem) *Let* $\{Y_\beta \subset \mathcal{R}^n : 0 \le \beta \le 1\}$ *be a family of compact subsets satisfying*

$Y_\beta \in \mathcal{K}^n$ *for all* $0 \le \beta \le 1$;

$Y_\beta \subseteq Y_\alpha$ *for* $0 \le \alpha \le \beta \le 1$;

$Y_\beta = \bigcap_{i=1}^\infty Y_{\beta_i}$ *for any nondecreasing sequence* $\beta_i \to \beta$ *in* $[0, 1]$.

Then there is a fuzzy set $u \in \mathcal{D}^n$ *such that* $[u]^\beta = Y_\beta$. *In particular, if the* Y_β *are also convex, then* $u \in \mathcal{E}^n$. *Conversely, the level sets of any* $u \in \mathcal{E}^n$, $[u]^\beta$ *are convex and satisfy these conditions.*

This is Theorem 1.5.1 in a suitable form.

Lemma 6.2.1. *Let Ω be an open subset of $\mathcal{R} \times \mathcal{R}^n$ and suppose that G is a usc mapping from Ω to E^n. Define $F(\cdot, \cdot, \beta) : \mathcal{R}^{n+1} \to \mathcal{K}_C^n$ to be the mapping $(t, x) \mapsto [G(t, x)]^\beta$. Then $F(\cdot, \cdot, \beta)$ is usc on Ω.*

Proof. Let $\mathcal{B}$ be the unit ball in E^n, that is $\mathcal{B} = \{u \in E^n : d(\chi_{\{0\}}, u) \leq 1\}$. By definition, for each $(t_0, x_0) \in \Omega$ and each $\epsilon > 0$ there exists $\delta > 0$ such that $\|(t, x) - (t_0, x_0)\| < \delta$ implies that $G(t, x) \leq G(t_0, x_0) + \epsilon\mathcal{B}$. Here, $u \leq v$ for $u, v \in E^n$ means that $u(\xi) \leq v(\xi)$ for all $\xi \in \mathcal{R}^n$ and implies the level set containment $[u]^\beta \subseteq [v]^\beta$ for all $0 \leq \beta \leq 1$. If $w \in \mathcal{B}$, $d_H([w]^\beta, \{0\}) \leq 1$ for each β, that is $[w]^\beta \in \mathcal{B}$. Hence, $[G(t, x)]^\beta \subseteq [G(t_0, x_0)]^\beta + \epsilon\mathcal{B}$ and the result follows.

We can now prove our main results concerning the level sets of solutions.

Theorem 6.2.3. *Let $X_0 \in E^n$ and let Ω be an open set in $\mathcal{R} \times \mathcal{R}^n$ containing $\{0\} \times supp(X_0)$. Suppose that $G : \Omega \to E^n$ is usc and write $F(t, x; \beta) = [G(t, x)]^\beta \in \mathcal{K}_C^n$ for all $(t, x, \beta) \in \mathcal{R}^{n+1} \times [0, 1]$. Let the boundedness assumption, with constants b, M, T, hold for all $x_0 \in X_0$ and the inclusion*

$$x'(t) \in F(t, x; 0), \quad x(0) \in supp(X_0). \quad (6.2.5)$$

Then the attainable sets $\mathcal{A}_\beta(X_0, T)$, $\beta \in [0, 1]$, of the family of inclusions

$$x'_\beta(t) \in F(t, x_\beta; \beta), \quad x(0) \in [X_0]^\beta, \quad \beta \in [0, 1], \quad (6.2.6)$$

are the level sets of a fuzzy set $\mathcal{A}(X_0, T) \in \mathcal{D}^n$. The solution sets $S_\beta(X_0, T)$ of (6.2.6) are the level sets of a fuzzy set $S(X_0, T)$ defined on $\mathcal{Z}_T(\mathcal{R}^n)$.

Proof. First, $S_\beta(X_0, T)$ is well-defined and compact. By Lemma 6.2.1, for each β the velocity $F(t, x; \beta)$ is usc. Since $F(t, x; \beta) \subseteq F(t, x; 0)$ and $[X_0]^\beta \subseteq supp(X_0)$ for all $0 < \beta \leq 1$, the boundedness assumption holds for each of (6.2.6) since it is true for (6.2.5) and $S_\beta(X_0, T) = \bigcup_{x \in X_\beta} S_\beta(x, T)$ exists, where $X_\beta = [X_0]^\beta$. Let x_k be a sequence of solutions in $S_\beta(X_0, T)$. Then $x_k(0) = \xi_k \in X_\beta$ and $x_k(t) \in \xi_k + (b + tM)\mathcal{B} \subseteq \Omega$, again by the boundedness assumption, whence $\|x'_k(t)\| \leq M$. So the assumptions of the compactness theorem (see Aubin and Cellina [1]) hold and there is a subsequence $x_{k(j)}$ converging to a solution x in $C([0, T], \mathcal{R}^n)$ and $x'_{k(j)}$ converges weakly* to x' in $L^\infty([0, T], \mathcal{R}^n)$. Using the convergence theorem as in the proof of Theorem 6.2.1 shows that $x'(t) \in F(t, x(t); \beta)$ a.e. and thus $S_\beta(X_0, T)$ is sequentially compact and so compact because $\mathcal{Z}_T(\mathcal{R}^n)$ is metrizable.

Obviously $S_\beta(X_0, T)$ is decreasing in β, since both $F(t, x; \beta)$ and X_β on the right of (6.2.6) are decreasing. Finally, an argument similar to that of Theorem 6.2.1 shows that for any nondecreasing sequence $\beta_n \uparrow \beta$, $\bigcap_n S_{\beta_n}(X_0, T) = S_\beta(X_0, T)$. The stacking theorem easily generalizes from $\mathcal{R}^n$ as base space to any Banach space and hence the S_β are level sets of a fuzzy set on $\mathcal{Z}_T(\mathcal{R}^n)$. The result for the $\mathcal{A}_\beta$ follows from this, and the theorem is proved.

Remark 6.2.1. If the condition that G be bounded on $[0, \infty) \times \Gamma$, where $\Gamma \subseteq \mathcal{R}^n$ is open, is added to the conditions of the theorem, the interval of existence and consequences extend to $[0, \infty)$.

We shall next prove a result on the boundedness of the solution set of (6.2.5).

Assume that $G : \mathcal{R}^+ \times \mathcal{R}^n \to E^n$ satisfies the conditions

(i) G is continuous;

(ii) there exists a real integrable function $k : \mathcal{R}^+ \to \mathcal{R}^+$ such that for all $x, y \in \mathcal{R}^n$, $t \in \mathcal{R}^+$,

$$d(G(t, x), G(t, y)) \le k(t) \|x - y\|,$$

$$d(\chi_{\{0\}}, G(t, 0)) \le k(t).$$

Theorem 6.2.4. *Under the assumptions* (i), (ii), *together with* $\int_0^\infty k(s)ds < \infty$, *the solution set* $S(X_0, t)$ *defined by the family of differential inclusions* (6.2.6) *is bounded for all time.*

Proof.   Consider each level set $S_\beta(X_0, t)$ of the solution, $0 \le \beta \le 1$. From a result on Constantin [11], the map S_β is continuous from $\mathcal{R}^n$ into the family of nonempty closed subsets of the Banach space of continuous and bounded functions from $\mathcal{R}^+$ to $\mathcal{R}^n$ endowed with the Hausdorff metric induced by the sup norm. Since X_0 has compact support, $S_\beta(X_0, t)$ is thus bounded for all $t \in \mathcal{R}^+$ and hence also must be $S(X_0, t)$, since $S_0(X_0, t) \supseteq S_\beta(X_0, t)$ is bounded.

6.3   Differential Inclusions

Let us describe the necessary results of the theory of differential inclusions in this section, so that we can formulate the theory of fuzzy differential inclusions in the next section. We shall repeat some notation and assumptions for convenience.

The most convenient solutions of differential inclusions are absolutely continuous. Recall that a continuous function $x : [0, T] \to Y \subseteq \mathcal{R}^n$ is said to be *absolutely continuous* if there exists a locally integrable function v such that

$$\int_t^s v(r)dr = x(s) - x(t)$$

for all $t, s \in [0, T]$. Then $x(\cdot)$ is defined to have derivative $x'(t) = v(t)$ almost everywhere (a.e.) in $[0, T]$. Let $\Omega \subset \mathcal{R} \times \mathcal{R}^n$ be an open subset containing $(0, x_0)$ and let $H : \Omega \to \mathcal{K}_C^n$. The differential inclusion

$$x'(t) \in H(t, x(t)) \quad \text{a.e.}, \quad x(0) = x_0, \tag{6.3.1}$$

is said to have a solution $y(t)$ on $[0, T]$ if $y(\cdot)$ is absolutely continuous, $y(0) = x_0$ and $y(\cdot)$ satisfies the inclusion a.e. in $[0, T]$. For simplicity, the epithet a.e. will be understood as present in all inclusions and not explicitly stated time after time.

The *boundedness assumption* is said to hold if there exist b, T, $M > 0$ such that

(i) the set $Q = [0, T] \times (x_0 + (b + MT)B^n) \subset \Omega$, where B^n is the unit ball of $\mathcal{R}^n$;

(ii) H maps Q into the ball of radius M.

Denote the set of all solutions of (6.3.1) on $[0, \tau]$ by $S(x_0, \tau)$ and the attainable set by $\mathcal{A}(x_0, \tau) = \{x(\tau) : x(\cdot) \in S(x_0, \tau)\}$ and write $\mathcal{Z}_T(\mathcal{R}^n) = \{x(\cdot) \in C([0, T]; \mathcal{R}^n) : x'(\cdot) \in L^\infty([0, T]; \mathcal{R}^n)\}$. It is known that for every $x_1 \in x_0 + b\,\text{int}\,B^n$, $\Sigma(x_1, T)$ exists and is a compact subset of $\mathcal{Z}_T(\mathcal{R}^n)$, and each attainable section $\mathcal{A}(x_1, \tau)$, $0 < \tau \leq T$, is a compact subset of $\mathcal{R}^n$, (see Aubin and Cellina [1]). In fact, although these sets are not in general convex, they are acyclic which is stronger than simply connected. See De Blasi and Myjak [15]. Write $\mathcal{A}(W, \tau) = \bigcup_{w \in W} \mathcal{A}(w, \tau)$.

Let $K \subseteq \mathcal{R}^n$. Solutions $x(t)$, $t \in J = [0, \tau]$, of a differential inclusion

$$x'(t) \in G(t, x), \quad x(0) = x_0 \in K,$$

are said to be *viable* if $x(t) \in K$ for all $t \in J$. In discussing periodicity, the interval J is finite, but for stability considerations, $J = [0, \infty)$.

It turns out that, under mild conditions on G, a tangency condition is necessary and sufficient for the existence of viable solutions on J, see Deimling [17] and Aubin and Cellina [1]. This tangency condition is expressed in terms of the *contingent cone* at $x \in K$

$$T_K(x) = \{y : \liminf_{h \to 0+} h^{-1}\rho(x + hy, K) = 0\}.$$

If $K \subset \mathcal{R}^n$ is a nonempty closed convex set, then $T_K(x)$, $x \in K$, is convex and

$$T_K(x) = \overline{\{\lambda(y - x) : \lambda \geq 0, y \in K\}},$$

which gives an idea of what the contingent cone looks like.

The main result that will be required on periodicity is the following result which is taken from Deimling [17].

Theorem 6.3.1. *Let $K \subset \mathcal{R}^n$ be a nonempty compact convex subset, and suppose that $G : \mathcal{R}^+ \times K \to \mathcal{K}_C^n$ is usc, that G be ω-periodic in t, $G(t+\omega, x) = G(t, x)$, $x \in K$, and that $\|G(t, x)\| \leq c(t)(1 + \|x\|)$ on $J \times K$, $J = [0, \omega]$, where $c \in L^1(J)$. If*

$$G(t, x) \bigcap T_K(x) \neq \emptyset \quad on \quad J \times K,$$

then $u'(t) \in G(t, u(t))$ has at least one ω-periodic solution.

Let $K \subset \mathcal{R}^n$ be nonempty and suppose that $G : \mathcal{R}^+ \times K \to \mathcal{K}_C^n$ be such that the initial value problems

$$x' \in G(t, x(t)), \ t_0 \leq t < \infty, \ x(t_0) = x_0, \tag{6.3.2}$$

have solutions for every $t_0 \geq 0$ and $x_0 \in K$. So, the interval of existence of solutions is $J = [0, \infty)$.

A set M is *stable* for the inclusion (6.3.2) if for all $\epsilon > 0$ and $t_0 \geq 0$ there exists $\delta = \delta(\epsilon, t_0) > 0$ such that $x_0 \in M + \delta B^n$ implies that $x(t) \in M + \epsilon B^n$ on $[t_0, \infty)$ for every solution $x(t)$ of (6.3.2). If $\mathcal{A}(x_0, t)$ is the attainability set of (6.3.2), this may be rephrased as $x_0 \in M + \delta B^n$ implies that $\rho(\mathcal{A}(x_0, t), M) \leq \epsilon$ on $[t_0, \infty)$. If $\delta = \delta(\epsilon)$ is independent of t_0 and depends only on ϵ, M (6.3.2) is said to be *uniformly stable* for the inclusion. If $\rho(\mathcal{A}(x_0, t), M) \to 0$ as $t \to \infty$ and M is (uniformly) stable, the set M is said to be (uniformly) *asymptotically stable*.

Example 6.3.1. *Consider the equation $y'' + \alpha y' + \mu sgn(y') + y = \sin t$, where α, $\mu > 0$, models dry friction. Here, $sgn(y) = y/|y|$, $y \neq 0$. This can be replaced by the inclusion*

$$x' \in G(t, x) = \begin{pmatrix} 0 & 1 \\ -1 & -\alpha \end{pmatrix} x + \begin{pmatrix} 0 \\ \sin t - \mu Sgn(x_2) \end{pmatrix}, \tag{6.3.3}$$

where the multivalued function $Sgn(y) = sgn(y)$, $y \neq 0$, but is $[-1, 1]$ when $y = 0$. Clearly, G satisfies a linear growth condition and is 2π-periodic, so *by Theorem 6.3.1 all solutions exist on* $[0, \infty)$. *If* x, y *are solutions, writing* $v(t) = \frac{1}{2}\|x(t) - y(t)\|^2$ *gives*

$$v' \leq -\alpha(x_2 - y_2)^2 - \mu(x_2 - y_2)(\bar{x}_2 - \bar{y}_2) \leq -\alpha(x_2 - y_2)^2,$$

with $\bar{x}_2 \in Sgn(x_w(t))$ $\bar{y}_2 \in Sgn(y_2(t))$. *So* $v'(t) \leq 0$ *and* $v(t) \leq v(t_0)$ *for* $t \geq t_0$. *Consequently, every solution* $x(t)$ *is uniformly stable. It can be shown that if* $\mu \geq 1$, *the only periodic solutions are constant,* $x(t) \equiv x_0$, *while if* $0 < \mu < 1$ *there is a unique nonconstant* 2π-*periodic solution* $p(t)$, *Deimling* [17]. *Clearly, none of the constant solutions nor the periodic solution are asymptotically stable, but certain sets are. For example,* $p(t+\tau)$ *is a solution which cannot approach* $p(t)$ *in the Lyapunov sense, but the set* $P = \{p(t) : t \geq 0\}$ *is asymptotically stable, which can be shown by Lyapunov theory.*

A function $V : \mathcal{R}^+ \times K \to \mathcal{R}+$ is a *Lyapunov function* of the inclusion (6.3.2) on K if

(1) $V(t, x) \geq \varphi(\|x\|)$ on $K \bigcap (rB^n)$, for some $r > 0$ and some continuous strictly increasing $\varphi : [0, r) \to \mathcal{R}^+$.

(2)

$$\frac{\partial}{\partial t} V(t, x) + \nabla V(t, x)u \leq -W(x) \leq 0 \qquad (6.3.4)$$

for all $t \geq t_0$, all $x \in K$ and every $u \in G(t, x)$, where $W : K \to \mathcal{R}^+$ is continuous on K.

If V is a Lyapunov function for (6.3.2) on K, define

$$E = \{x : W(x) = 0, x \in K\}.$$

In particular, (6.3.4) implies that a Lyapunov function is nonincreasing on the solution set of (6.3.2) as time evolves. So, if $K_0 \subseteq K$ is compact and $V(t_1, x) \leq a$ for all $x \in K_0$ and some $t_1 \geq t_0$, then solutions which start in K_0 remain in K_0 and are thus bounded. If $x(t)$ is a trajectory of the inclusion (6.3.2) and $K_1 \subset K$, if $\rho(x(t)) \to K_1$ as $t \to \infty$, write $x(t) \to K_1$, $t \to \infty$.

Theorem 6.3.2. *Let* V *be a Lyapunov function for (6.3.2) on* K, *and suppose that the attainability set* $\mathcal{A}(x_0, t)$ *remains in* K *for all* $t_0 \leq t < \infty$. *If for each solution* $x(t)$, $W(x(t))$ *is absolutely continuous and its derivative is bounded above a.e. on* $[t_0, \infty)$, *then* $x(t) \to E$ *as* $t \to \infty$. *In particular, if* $V = V(x)$ *does not explicitly depend on* t, *then* $x(t) \to E(c) :$ $= E \bigcap \{x : V(x) \leq c\}$ *for some* $c > 0$.

For a proof, see LaSalle [71].

As an example, let $r(t) = [r_1(t), r_2(t)]$, $0 \leq t < \infty)$, be an interval valued function, with $r_1(t) \geq \gamma > 0$ bounded away from 0 for all $t \geq 0$. Consider the two-dimensional inclusion

$$x' \in \begin{pmatrix} 0 & 1 \\ -1 & -r(t) \end{pmatrix} x. \qquad (6.3.5)$$

Let $V(x) = x_1^2 + x_2^2$. Then for every $r_0 \in r(t)$, checking (6.3.4) gives

$$\nabla V(x)u = x_1 x_2 + x_2(-r_0 x_2 - x_1) = -r_0 x_2^2 \leq -\gamma x_2^2,$$

so $W(x) = \gamma x_2^2$, E is the x_1-axis and $x_2(t) \to 0$ as $t \to \infty$. But, since V is independent of t, $x_1(t) \to$ a bounded subset of E as $t \to \infty$.

6.4   Fuzzy Differential Inclusions

Having the necessary results on differential inclusions in Sections 6.2 and 6.3, we are now ready to investigate fuzzy differential inclusions.

Let $H : \mathcal{R} \times \mathcal{R}^n \to E^n$ and consider the fuzzy differential equation (FDE)

$$x' = H(t, x), \quad x(0) = X_0 \in E^n,$$

interpreted as a family of differential inclusions. Set $[H(t, x)]^\beta = F(t, x; \beta)$ and identify the FDE with the family of differential inclusions

$$x'_\beta(t) \in F(t, x_\beta(t); \beta) \; x_\beta(0) = x_0 \in [X_0]^\beta, \; 0 \leq \beta \leq 1. \qquad (6.4.1)$$

where Ω is an open subset of $\mathcal{R}^{n+1}$ containing $(0, [X_0]^0)$, $\beta \in I := [0, 1]$ and $F : \Omega \times I \to \mathcal{K}_C^n$. The boundedness assumption now holds if the set Q is as above and F maps $Q \times I$ into the ball of radius M. Denote the set of all solutions of (6.4.1) on $[0, \tau]$ by $S_\beta(x_0, \tau)$ and the attainable set by $\mathcal{A}_\beta(x_0, \tau) = \{x(\tau) : x(\cdot) \in S_\beta(x_0, \tau)\}$. As seen before, $S_\beta(x_1, T)$ exists and is a compact subset of $\mathcal{Z}_T(\mathcal{R}^n)$, and each attainable section $\mathcal{A}_\beta(x_1, \tau)$, $0 < \tau \leq T$, is a compact subset of $\mathcal{R}^n$.

The results on periodicity and stability for differential inclusions were in the space $\mathcal{R}^n$. When extending these ideas to FDEs, the definitions of stability have to be formulated in $\mathcal{D}^n$, and notions of periodicity for solution sets are also in different spaces. If $U \in \mathcal{D}^n$ is a fuzzy set and $\mathcal{U}, \mathcal{W} \subset \mathcal{D}^n$ are closed subsets of $\mathcal{D}^n$, define the distance from $\mathcal{W}$ and Hausdorff separation respectively by

$$\rho_*(U, \mathcal{W}) = \inf_{W \in \mathcal{W}} d(U, W)$$

$$\rho_D(\mathcal{U}, \mathcal{W}) = \sup_{U \in \mathcal{U}} \rho_*(U, \mathcal{W}).$$

The significance of these definitions is that, in the metric space $(\mathcal{D}^n, d)$ of fuzzy sets, $\rho_*(U, \mathcal{W})$ is the distance of $U \in \mathcal{D}^n$ from $\mathcal{W} \subset \mathcal{D}^n$ and is the analog of $\rho(x, A)$ in $\mathcal{R}^n$. Correspondingly, $\rho_D(\mathcal{U}, \mathcal{W})$ is the Hausdorff separation between $\mathcal{U}, \mathcal{W} \subset \mathcal{D}^n$ with respect to the metric d and is the analog of the Hausdorff separation $\rho(A, B)$ in $\mathcal{K}^n$.

Let $\mathbf{0}$ be the fuzzy singleton $\chi_{\{0\}} \in \mathcal{D}^n$, write $\|U\| = d(U, \mathbf{0})$ and denote the open unit ball in $\mathcal{D}^n$ by $\mathcal{B}^n = \{U \in \mathcal{D}^n : \|U\| < 1\}$.

A set $\mathcal{U} \subset \mathcal{D}^n$ is *stable* for the FDE (6.4.1) if for all $\epsilon > 0$ and $t_0 \geq 0$ there exists $\delta = \delta(\epsilon, t_0)$ such that $X_0 \in \mathcal{U} + \delta\mathcal{B}^n$ implies that $\mathcal{A}(X_0, t) \in \mathcal{U} + \epsilon\mathcal{B}^n$ on $[t_0, \infty)$, where $\mathcal{A}(X_0, t)$ is the fuzzy attainability set defined by the family (6.4.1). That is, $\rho_*(X_0, \mathcal{U}) < \delta$ implies that $\rho_D(\mathcal{A}(X_0, t), \mathcal{U}) \leq \epsilon$ on $[t_0, \infty)$. If $\delta = \delta(\epsilon)$ is independent of t_0 and depends only on ϵ, $\mathcal{U}$ for the FDE (6.4.1) is said to be *uniformly stable*. If $\rho_D(\mathcal{A}(X_0, t), \mathcal{U}) \to 0$ as $t \to \infty$ and $\mathcal{U}$ is (uniformly) stable, the set $\mathcal{U}$ is said to be (uniformly) asymptotically stable. Most frequently, $\mathcal{U}$ will consist of a single fuzzy set $U \in \mathcal{D}^n$.

Suppose that $\mathcal{U} \subset \mathcal{D}^n$ is support bounded, that is

$$K = \text{supp}(\mathcal{U}) := \bigcup_{u \in \mathcal{U}} \text{supp}(u)$$

is a bounded set in $\mathcal{R}^n$. A function $V : \mathcal{R}^+ \times K \to \mathcal{R}^+$ is a Lyapunov function for the FE (6.4.1) if V is a Lyapunov function for the differential inclusion

$$x'(t) \in F(t, x; 0), \quad x(0) \in \text{supp}(X_0), \tag{6.4.2}$$

for all $X_0 \in \mathcal{U}$.

Recall that any crisp set $X \subset \mathcal{R}^n$ is also a fuzzy set, the membership function being the characteristic function χ_X.

Theorem 6.4.1. *Let V be a Lyapunov function for (6.4.1) on $K = \text{supp}(\mathcal{U})$, and suppose that the attainability set $\mathcal{A}(x_0, t)$ remains in $\mathcal{U}$ for all $t_0 \leq t < \infty$. If for each solution $x(t)$ of (6.4.2), $W(x(t))$ is absolutely continuous and its derivative is bounded above a.e. on $[t_0, \infty)$, then $\rho_D(\mathcal{A}(x_0, t), \chi_E) \to 0$ as $t \to \infty$. In particular, if $V = V(x)$ does not explicitly depend on t, then writing $E(c) = E \bigcap \{x : V(x) \leq c\}$, $\rho_D(\mathcal{A}(x_0, t), \chi_{E(c)}) \to 0$ for some $c > 0$ as $t \to \infty$.*

Proof. From Theorem 6.3.2 as $t \to \infty$ each solution $x(t)$ of (6.4.2) approaches E, hence the attainability set $\mathcal{A}(x_0, t)$ satisfies $\rho(\mathcal{A}(x_0, t), E)$

$\rightarrow 0$ as $t \rightarrow \infty$. By Theorem 6.2.3, the FDE (6.4.1) has fuzzy attainability set $\mathcal{A}(x_0, t)$ and $\mathcal{A}(x_0, t) = [\mathcal{A}(x_0, t)]^0$. Since $[\mathcal{A}(x_0, t)]^\beta \subseteq \mathcal{A}(x_0, t)$, $0 < \beta \leq 1$, it follows that $\rho([\mathcal{A}(x_0, t)]^\beta, E) \rightarrow 0$ as $t \rightarrow \infty$. Hence, $\rho_D(\mathcal{A}(x_0, t), \chi_E) \rightarrow 0$ as $t \rightarrow \infty$. Again by Theorem 6.3.2, if V does not depend on t, $\rho(\mathcal{A}(x_0, t), E(c)) \rightarrow 0$ for some c as $t \rightarrow \infty$. The second part of the result now proceeds by the same reasoning as above.

Example 6.4.1. *To illustrate the theorem, fuzzify the differential equation*

$$x' = \begin{pmatrix} 0 & 1 \\ -1 & -r(t) \end{pmatrix} x \qquad (6.4.3)$$

by letting $r(t)$ be a symmetric triangular valued fuzzy function with level sets $[r(t)]^\beta = (1 - \beta)[r_1(t), r_2(t)] := (1 - \beta)R(t)$ and with $r_1(t) \geq \gamma > 0$. This gives the family of inclusions

$$x'_\beta \in \begin{pmatrix} 0 & 1 \\ -1 & -R(t)(1 - \beta) \end{pmatrix} x_\beta, \quad 0 \leq \beta \leq 1. \qquad (6.4.4)$$

Then, referring to (6.3.5), $V(x) = x_1^2 + x_2^2$ is a Lyapunov function for the FDE (6.4.3) and $\rho_D(\mathcal{A}(x_0, t), \chi_{E(c) \times \{0\}}) \rightarrow 0$ as $t \rightarrow \infty$, where $E(c)$ is the attractive set approached by $x_1(t)$ in (6.3.5).

Example 6.4.2. *In Example 6.3.1, suppose that the frictional parameter μ is known only vaguely and that the term $F(x') = \mu Sgn(x')$, $\mu \geq 1$, is modeled by the trapezoidal fuzzy number valued function whose β-levels are given by*

$$[F(z)]^\beta = \begin{cases} \mu(1 - \beta)Sgn(z)[1, 2] & \text{if } z \neq 0, \\ \mu(2 - \beta)[-1, 1] & \text{if } z = 0. \end{cases} \qquad (6.4.5)$$

Instead of (6.3.3), consider now the corresponding FDE represented by the family of inclusions

$$x' \in [G(t, x)]^\beta = \begin{pmatrix} 0 & 1 \\ -1 & -\alpha \end{pmatrix} x + \begin{pmatrix} 0 \\ sint - [F(x_2)]^\beta \end{pmatrix}, \qquad (6.4.6)$$

where $0 \leq \beta \leq 1$. Here the level set (6.4.5) replaces the term $\mu Sgn(x_2)$. Clearly, $[G(t, x)]^\beta$ satisfies the conditions of Theorem 5.3.1, so periodic solutions to the β-th inclusion exist.

Adapting an argument of Deimling [17], fix β and suppose that $x(t) \equiv x^*$ is a constant solution for the β-inclusion of (6.4.6). Since the inclusion is just

$$x'_1 = x_2$$

$$x'_2 \in -x_1 - \alpha x_2 + \sin t - [F(x_2)]^\beta,$$

$x^*_2 = 0$, *while* $-x^*_1 + \sin t - \mu w(t) = 0$, *for some* $w(t) \in [-(2 - \beta), 2 - \beta]$. *Hence,*

$$|\sin t - x^*_1| \leq \mu|w(t)| \leq \mu(2 - \beta), \tag{6.4.7}$$

and consequently

$$1 - \mu(2 - \beta) \leq x^*_1 \leq -1 + \mu(2 - \beta). \tag{6.4.8}$$

This last inequality (6.4.8) *implies that* $\mu \geq 1/(2 - \beta)$. *In particular, if* $x(t) \equiv x^*$ *is a constant solution for each inclusion of the family,* $\mu \geq 1$.

On the other hand, if $x(t)$ *is a* 2π-*periodic solution, then so also is* $y(t) = x(t) - x^*$ *for any* $x^* = (x^*_1, 0)^T$ *with* x^*_1 *satisfying* (6.4.8). *So,*

$$y'_1 = y_2, y'_2 = -y_1 - x^*_1 - \alpha y_2 + \sin t - \mu w(t),$$

with $w(t) \in (2 - \beta)Sgn(y_2) \subseteq [-(2 - \beta, 2 - \beta]$. *From the first equation*

$$0 = [y_1(t)^2/2]^{2\pi}_0 = \int_0^{2\pi} y'_1(t)y_1(t)dt = \int_0^{2\pi} y_1(t)y_2(t)dt$$

and using this in the second equation gives

$$
\begin{aligned}
0 &= \int_0^{2\pi} y'_2 y_2 dt \\
&= \int_0^{2\pi} y_1 y_2 dt - \alpha \int_0^{2\pi} y_2(t)^2 dt + \int_0^{2\pi} (\sin t - x^*_1 - \mu w(t))y_2 dt \\
&= -\alpha \int_0^{2\pi} y_2(t)^2 dt + \int_0^{2\pi} ((\sin t - x^*_1)sgn(y_2) - \mu w(t)|y_2|dt \\
&\leq -\alpha \int_0^{2\pi} y_2^2 dt,
\end{aligned}
$$

since $|\sin t - x^*_1| \leq \mu|w(t)|$ *and* $sgn(y_2)y_2 = |y_2|$. *Thus,* $y_2(t) = 0$ *a.e. and so* $y'_1(t) = 0$. *It follows that* $x_2(t) = 0$ *and* $x_1(t) = y_1(t) + x^*_1$ *is constant. Consequently, if* $\mu \geq 1$ *the only* 2π-*periodic solutions are* $x(t) = (x^*_1, 0)$, *with* x^*_1 *satisfying* $1 - \mu \leq x^*_1 \leq -1 + \mu$. *Moreover, if* $\mu \geq 1/(2 - \beta)$, *the only* 2π-*periodic solutions of the* β-*th inclusion are* $x(t, \beta) = (x^*_1(\beta), 0)$ *with* $x^*_1(\beta)$ *satisfying* (6.4.8). *Now, using the Lyapunov function* $V = (x_1 - x^*_1)^2/2 + x_2^2/2$ *gives, for* $w(t) \in [F(x_2)]^0$,

$$
\begin{aligned}
V' &= (x_1 - x^*_1)x_2 + x_2(-x_1 - \alpha x_2 + \sin t - \mu w(t) \\
&= -\alpha x_2^2 + x_2(-x^*_1 + \sin t - \mu w(t)) \\
&= -\alpha x_2^2 + |x_2|((\sin t - x^*_1)sgn(x_2) - \mu w(t)) \\
&\leq -\alpha x_2^2 = W(x),
\end{aligned}
$$

from (6.4.7). Hence, the attainability set $\mathcal{A}(X_0, t)$ is attracted to $U \in \mathcal{D}^2$ given by

$$[U]^\beta = [1 - \mu(2 - \beta), -1 + \mu(2 - \beta)] \times \{0\}.$$

Let us next consider the periodicity of solutions. Under appropriate conditions, such as those of Theorem 6.3.1, an inclusion $x' \in G(t, x)$ will have ω-periodic solutions. Denote by $\mathcal{A}_P(W, t)$, $0 \leq t \leq \omega$, the attainability set of all ω-periodic solutions $x(t)$ such that $x(0) \in W$. Clearly, $t \mapsto \mathcal{A}_P(W, t)$ is an ω-periodic set-valued function on $0 \leq t < \infty$.

An FDE of the form (6.4.1) is said to have an ω-periodic solution $\mathcal{A}_P$ (X_0, t) if there is a family of ω-periodic set-valued functions $t \mapsto U(t, \beta)$, $0 \leq \beta \leq 1$, such that $U(t, \beta) \subseteq \mathcal{A}([X_0]^\beta, t)$ for $t \geq 0$ and for each $t \geq 0$ the family $\{U(t, \beta)\}$ satisfies the conditions of the stacking theorem 6.2.2. That is, for all $t \geq 0$

$$[\mathcal{A}_P(X_0, t)]^\beta = U(t, \beta).$$

One can also speak of the mapping from $[0, \infty)$ to the function space $\mathcal{Z}_\infty(\mathcal{R}^n)$ given by $t \mapsto S_P(X_0, t)$ as being ω-periodic. However, the main interest for applications and computation is the periodic attainability set.

As an immediate consequence of Theorems 6.3.1 and 6.2.3, we have

Theorem 6.4.2. *Let $X_0 \in E^n$, $K \in \mathcal{K}^n$ and let $H : \mathcal{R}^+ \times K \to E^n$ be usc. Suppose that H is ω-periodic in t and that $\|H(t, x)\| \leq c(t)1 + \|x\|)$ on $J \times K$, $J = [0, \omega]$, $\int_0^\omega c(t)dt < \infty$. If $[H(t, x)]^1 \bigcap T_K(t) \neq \emptyset$ on $J \times K$, then there exists a nonempty ω-periodic fuzzy attainability set $\mathcal{A}_P(X_0, t)$ and a nonempty ω-periodic fuzzy solution set $S(X_0, t)$.*

Note that the contingency condition of Theorem 6.3.1 is satisfied at every β-level, since $[H(t, x)]^\beta \supseteq [H(t, x)]^1$, while the usc of $[H(t, x)]^\beta$ follows from that of $H(t, x)$.

Example 6.4.3. *Again fuzzyifying Example 6.3.1, modeling $F(z) = \mu Sgn(z)$ by (6.4.5), but with $0 \leq \mu < 1$, as before, $[G(t, x)]^\beta$ in (6.4.6) satisfies Theorem 6.3.1 and 2π-periodic solutions to the β-th inclusion exist for $0 \leq \beta \leq 1$. Since $0 \leq \mu < 1$, Example 6.4.2 shows that there are nonconstant 2π-periodic solutions at each β-level. Suppose that $x(t)$, $y(t)$ are such solutions. Using th Lyapunov function $V(t) = \|x(t) - y(t)\|^2/2$ gives, as in Example 6.3.1, $V'(t) \leq -\alpha(x_2 - y_2)^2$ and so on $J = [0, 2\pi]$, by periodicity,*

$$V(2\pi) \leq V(t) \leq V(0) = V(2\pi).$$

Hence, $x_2(t) = y_2(t)$, which from (6.4.6) implies that $x_1' = x_2 = y_2 = y_1'$ and thus $x(t) = y(t) + (x^*, 0)$ on J. Since $x_2' = y_2'$, from the inclusion (6.4.6) *it follows that* $[F(x_2)]^\beta \bigcap (x^* + [F(y_2)]^\beta) \neq \emptyset$. *Hence,* $|x^*| \leq$ diam$([F(x_2)]^\beta)$ *and so*

$$
\begin{aligned}
\text{diam}([\mathcal{A}(X_0, t)]^\beta) &= \sup\{\|x(t) - y(t)\| : x(t), y(t) \in \mathcal{A}(X_0, t)]^\beta \\
&\leq \text{diam}([F(z)]^\beta) \\
&\leq \mu(4 - 2\beta),
\end{aligned}
$$

from (6.4.5). Hence, the trapezoidal fuzzy number U, *given by* $[U]^\beta = \mu(2 - \beta)[-1, 1]$, $0 \leq \beta \leq 1$, *is an absorbing set for the attainability set* $\mathcal{A}(X_0, t)$ *as* $t \to \infty$. *Consequently, from Diamond [18] there exists a cocycle attractor for the FDE (6.4.6).*

Example 6.4.4. *Consider the FDE on* $\mathcal{R}^+$,

$$
y'(t) = -y(t) + W \cos t,
$$

where W *is the symmetric trapezoidal fuzzy number with level sets*

$$
[W]^\beta = \left[e_1 + \frac{1 - \beta}{2}(e_2 - e_1 - e), e_2 - \frac{1 - \beta}{2}(e_2 - e_1 - e)\right],
$$

where $e_1 < e < e_2$. *Interpreting the FDE as a family of inclusions*

$$
x'_\beta(t) \in -x_\beta(t) + \cos t [W]^\beta, \quad 0 \leq \beta \leq 1, \tag{6.4.9}
$$

clearly every solution of the form $x_\beta(t) = \frac{c_\beta}{2}(\cos t + \sin t)$, $c_\beta \in [W]^\beta$, *is a periodic solution for all* β, *since*

$$
\begin{aligned}
x'_\beta(t) &= -\frac{c_\beta}{2}(\cos t + \sin t) + c_\beta \cos t \\
&\in -x_\beta(t) + \cos t [W]^\beta.
\end{aligned}
$$

In fact, the solution set is given for $t \geq 0$ *and* $x(0) \in X_0$ *by*

$$
\begin{aligned}
x_\beta(t) &\in \frac{1}{2}(\sin t + \cos t)[W]^\beta + \left([X_0]^\beta - \frac{1}{2}[W]^\beta\right)e^{-t} \\
&= [S(t, 0, X_0)]^\beta.
\end{aligned}
$$

Clearly, the attainability set $\mathcal{A}(X_0, t)$ *approaches the fuzzy set* $E = \{W(\sin t + \cos t)/2 : t \geq 0\} = \sqrt{2}W/2$.

6.5   The Variation of Constants Formula

Recall the classical variation of constants formula for a nonhomogeneous n-dimensional system of first-order differential equations (DEs)

$$x'(t) = A(t)x(t) + f(t), \quad x(0) = x_0,$$

for example, the control system $x' = A(t)x + B(t)u(t)$. The solution may be written as

$$x(t) = \Phi(t)x_0 + \int_0^t \Phi(t)\Phi(s)^{-1}f(s)ds, \tag{6.5.1}$$

where $\Phi(t,s) = \Phi(t)\Phi(s)^{-1}$ is the *state transition matrix* and $\Phi(t)$ satisfies the matrix DE

$$\Phi'(t) = A(t)\Phi(t), \quad \Phi(0) = I. \tag{6.5.2}$$

In the special case when A is time-independent, $\Phi(t) = e^{tA}$ involves the matrix exponential and $\Phi(t,s) = \Phi(t-s) = e^{(t-s)A}$.

Let E^n be the space of all normal, upper semicontinuous and fuzzy convex sets, with compact support, on $\mathcal{R}^n$, endowed with the sup metric on level sets d as before. Denote by $\mathcal{M}^n$ the space of $n \times n$ matrices with entries in $\mathcal{E}^1$, that is, whose entries are fuzzy numbers. If $A \in \mathcal{M}^n$, it also lies in $\mathcal{E}^{n \times n}$. Consequently, $A : [0,T] \to \mathcal{M}^n$ is said to be continuous if it is continuous in the ϵ, δ sense in the $|\cdot|$ and d metrics.

We interpret (6.5.2) as the family of differential inclusions

$$\phi'_\beta(t) \in A_\beta(t)\phi_\beta(t), \quad \phi_\beta(0) = I, \quad 0 \le \beta \le 1, \tag{6.5.3}$$

where $A_\beta(t)$ denotes the level set of $A(t)$. That is, if $U(\cdot)$ is a measurable matrix valued selection in the set of matrices $A_\beta(\cdot)$, $\phi'_\beta(t) = U(t)\phi_\beta(t)$ a.e. and $\phi_\beta(0) = I$. Such solutions are guaranteed by Carathéodory's theorem, over an interval $[0,T]$, $0 < T \le \infty$, provided the end-point matrix valued functions $\underline{A}_\beta$, $\bar{A}_\beta$ are majorized by an integrable function on $[0,T]$. Denote

$$\Phi_\beta(t) = \{Y(t) : Y' = U(t)Y, Y(0) = I, U(\cdot) \in A_\beta(\cdot)\}.$$

Clearly, $\Phi_\beta(t)$ is nonempty. By the results of Section 6.2, the $\Phi_\beta(\cdot)$ are the level sets of the $\mathcal{D}^{n \times n}$ valued fuzzy function $\Phi(\cdot)$.

In particular, if A is a constant fuzzy matrix and the FDE is time-independent, define

$$\Phi_\beta = \{Y(t) : Y' = U(t)Y, Y(0) = I\}$$

for measurable matrix valued selections $U(\cdot) \in A_\beta$. Then, the $Y(t)$ are found in the usual way: find a basis of vector solutions $u_1(t), u_2(t), \ldots, u_n(t)$ for the differential equation $y' = U(t)y$ and form the matrix $Z(t = [u_1 \; u_2 \; \cdots \; u_n]$. Put $Y(t) = Z(t)Z(0)^{-1}$.

The $\Phi_\beta(t)$ now form the level sets of an $E^{n \times n}$-valued function, since an $n \times n$ interval matrix is a convex set in $\mathcal{R}^{n \times n}$. This has outlined a proof of the following result.

Theorem 6.5.1. *Let $A : [0, T] \to E^{n \times n}$ be a continuous fuzzy matrix valued function. Then (6.5.3) has a solution set $\Phi(t)$ which is a $\mathcal{D}^{n \times n}$ valued fuzzy matrix function. If A is constant, $\Phi(t)$ is an $E^{n \times n}$ valued function whose level sets are interval matrices.*

Remarks 6.5.1.

(1) Note that $\Phi(t)$ is invertible in the sense that each trajectory ϕ_β of the β-th inclusion (6.5.3) is invertible. Indeed, it is well known that $\psi_\beta = \phi_\beta^{-1}$ satisfies the matrix DE $\psi_\beta' = -\psi_\beta U(t)$ a.e., $\psi_\beta(0) = I$.

(2) If the initial time is t_0, the notation $\Phi(t, t_0)$ should be used, $\Phi(t_0, t_0) = I$. Here we have abbreviated $\Phi(t) := \Phi(t, 0)$. Note that $\Phi^{-1}(t, t_0) = \Phi(t_0, t)$. For constant A, $\Phi^{-1}(t) = \Phi(-t)$, sharing this property with the classical matrix exponential solution.

(3) Carathéodory's theorem is an existence result only and solutions may not be unique. However, there exist *maximum* and *minimum* solutions.

Since A_β is an *interval matrix*, that is, has compact real intervals as entries, U belongs to the interval $[\underline{A}_\beta, \bar{A}_\beta]$. Here, $\underline{A}$, $\bar{A}$ denote ordinary matrices whose elements are, respectively, the lower and upper end points of the real intervals (see, Neumaier [80] for notation and theory). This gives a method for evaluating $\Phi_\beta(\cdot)$ as an interval matrix.

In the case where $\underline{A}_\beta$ is a nonnegative matrix, that is, all elements of the matrix are nonnegative, or $\bar{A}_\beta$ is a nonpositive matrix, the computation is especially simple. If $B - A$ is a nonnegative matrix, write $B \geq A$.

Lemma 6.5.1. *Let A be a nonnegative matrix and suppose that $B \geq A$. If*

$$X'(t) = AX, \quad X(0) = I,$$

$$Y'(t) = BY, \quad Y(0) = I,$$

then $Y(t) \geq X(t)$, $t \geq 0$. In particular, if $A = [\underline{A}, \bar{A}]$ is an interval matrix with $0 \leq \underline{A} \leq U \leq \bar{A}$ and

$$\underline{X}'(t) = \underline{A}X(t), X'(t) = UX(t), \bar{X}'(t) = \bar{A}\bar{X}(t),$$

$\underline{X}(0) = X(0) = \bar{X}(0) = I$, *then* $\underline{X}(t) \leq X(t) \leq \bar{X}(t)$, $t \geq 0$.

Proof. This is a simple consequence of Theorem 5.1.1, pp. 315–316 of Lakshmikantham and Leela [61], since the matrix differential equations are equivalent to

$$X(t) = I + \int_0^t AX(s)ds,$$

$$Y(t) = I + \int_0^t BY(s)ds \geq I + \int_0^t AY(s)ds,$$

and the function $X \mapsto AX$ is monotonic nondecreasing in the partial order induced by the positive orthant.

Clearly, a similar results holds for nonpositive matrices. Indeed, it could be generalized to K-monotone matrices, where K is a cone, but for simplicity this is omitted. In the case where the interval matrix is not of these types, the interval matrix function $AX(t)$ will have extreme points corresponding to matrices internal to the interval matrix A. This can be estimated numerically by solving the matrix DEs on a gird, but this obviously significantly increases the computation involved.

Example 6.5.1. *As a numerical illustration, consider the system where $n = 2$ and A is given by*

$$A = \begin{pmatrix} (1;2)_s & (0.5;1.5)_s \\ (0.5;1.5)_s & (2;3)_s \end{pmatrix}.$$

Expressing the β-levels of A as an interval matrix,

$$A_\beta = \begin{pmatrix} (1+0.5\beta, 2-0.5\beta) & (0.5+0.5\beta, 1.5-0.5\beta) \\ (0.5+0.5\beta, 1.5-0.5\beta) & (2+0.5\beta, 3-0.5\beta) \end{pmatrix},$$

this interval system matrix is a positive family, because every matrix contained in the interval is a positive matrix and, applying the lemma, only the end point equations need be solved. So, for example, if $\beta = 0.0$, $\Phi_{0.0}(t) = [\underline{\Phi}_{0.0}(t), \bar{\Phi}_{0.0}(t)]$. Using the MATLAB function EIG for the eigenvalue-eigenvector calculations (INV is not needed because $\underline{A}$, $\bar{A}$ are symmetric here),

$$\underline{\Phi}_{0.0}(t) = \begin{pmatrix} 0.8536e^{\underline{\kappa}t} + 0.1465e^{\underline{\nu}t} & -0.3536e^{\underline{\kappa}t} + 0.3536^{\underline{\nu}t} \\ -0.3536e^{\underline{\kappa}t} + 0.3536e^{\underline{\nu}t} & -.1456e^{\underline{\kappa}t} + 0.8536e^{\underline{\nu}t} \end{pmatrix}, \quad (6.5.4)$$

where $\underline{\kappa} = 0.7929$, $\underline{\nu} = 2.2071$, and

$$\bar{\Phi}_{0.0}(t) = \begin{pmatrix} 0.6580e^{\bar{\kappa}t} + 0.3419e^{\bar{\nu}t} & -0.4743e^{\bar{\kappa}t} + 0.4743e^{\bar{\nu}t} \\ -04.4743e^{\bar{\kappa}t} + 0.4743e^{\bar{\nu}t} & 0.3419e^{\bar{\kappa}t} + 0.6580e^{\bar{\nu}t} \end{pmatrix}, \quad (6.5.5)$$

where $\bar{\kappa} = 0.9189$, $\bar{\nu} = 4.0811$.

We now proceed to discuss the variation of constants formula. In practice, the crisp fundamental matrix $\Phi(t, t_0)$ is difficult to find explicitly if the system matrix is time-dependent. Apart from theoretical discussions, the variation of constants formula is usually only useful when A is constant. For brevity, the discussion of this section will be for constant A, but the general case is not difficult and the more general proof is obvious.

Let $A \in E^{n \times n}$ be a matrix with fuzzy number entries, $X^0 \in E^n$ and let $F : [0, T] \to E^n$ be a continuous function (and hence strongly integrably bounded, with integrable selections). Consider the fuzzy initial value problem

$$x'(t) = Ax + F(t), \quad x(0) = X^0,$$

that is, for $0 \le \beta \le 1$, the family of inclusions

$$x'_\beta(t) \in A_\beta x_\beta(t) + F_\beta(t), \quad x_\beta(0) \in X^0_\beta. \quad (6.5.6)$$

Here, the suffixes β indicate level sets. Let $\Phi(t)$ indicate the solution set of the time-independent matrix FDE $\Phi' = A\Phi$, $\Phi(0) = I$. From the inclusion (6.5.6), there is a measurable matrix valued function $U(\cdot) \in A_\beta$, an integrable selection $f_\beta(\cdot) \in F_\beta(\cdot)$ and $y_0 \in X^0_\beta$ such that

$$x'_\beta(t) = U(t)x_\beta(t) + f_\beta(t) \text{ a.e.}, \quad x_\beta(0) = y_0.$$

Denote by $X_U(t)$ a solution to the equation $X'_U = U(t)X_U$, $X_U(0) = I$. From the classical formula (6.5.1),

$$x_\beta(t) = X_U(t)y_0 + \int_0^t X_U(t - s)f_\beta(s)ds. \quad (6.5.7)$$

Consequently, the solution set of (6.5.6) is

$$S_\beta(t) = \bigcup \{X_U(t)y_0 + \int_0^t X_U(t - s)f_\beta(s)ds : U \in A, f_\beta \in F_\beta, y_0 \in X^0_\beta\},$$

$$S_\beta(t) = \Phi_\beta(t)X^0_\beta + \int_0^t \Phi_\beta(t - s)F_\beta(s)ds, \quad (6.5.8)$$

by Theorem 6.5.1 and the definition of the Aumann integral [2], where the union has been taken over *all maximal and minimal* Carathéodory solutions, so as to give (6.5.8). Since for $0 \le t \le T$, $\Phi_\beta(t)$, X_β^0 and $F_\beta(t)$ are all compact, so also is $S_\beta(t)$. Moreover, clearly $S_\beta(t) \subseteq S_{\beta'}$ for $0 \le \beta' \le \beta \le 1$.

Theorem 6.5.2. *Let $A \in E^{n \times n}$ be a matrix with fuzzy number entries, $X^0 \in E^n$ and let $F : [0, T] \to E^n$ be a continuous function. The fuzzy initial value problem*

$$x'(t) = Ax + F(t), \quad x(0) = X^0,$$

has an absolutely continuous E^n valued solution $S(t)$, a.e. $t \in [0, T]$, given by the variation of constants formula

$$S(t) = \Phi(t)X^0 + \int_0^t \Phi(t - s)F(s)ds.$$

Here, $\Phi(t)$ is the fuzzy matrix valued solution of

$$\Phi'(t) = A\Phi(t), \quad \Phi(0) = I.$$

Proof.   From the comments above, to prove that the family $S_\beta(t)$ are the level sets of a fuzzy set $S(t)$ over $C([0]T, \mathcal{R}^n)$, it only remains to show upper semicontinuity (usc). That is, of $\beta_n \to \beta$ is a nondecreasing sequence, then $S_{\beta_n}(t) \to S_\beta(t)$. To see this, note that $\tilde{S}_\beta(t) = \bigcap_n \Sigma_{\beta_n}(t)$ is nonempty, since the intersection is of compact, nonempty nested sets. Clearly, $\tilde{S}_\beta(t) \supseteq S_\beta(t)$. But $X_{\beta_n}^0 \to X_\beta^0$ and $\Phi_{\beta_n}(t) \to \Phi_\beta(t)$ by usc. Let $x_{\beta_n} \in S_{\beta_n}$. Then there are $f_{\beta_n} \in F_{\beta_n}$, $U_n \in A_{\beta_n}$ and $y_n \in X_{\beta_n}^0$ satisfying (6.5.7). The sequence $\{y_n\}$ is bounded in $\mathcal{R}^n$ since X^0 has compact support and, similarly, U_n is bounded in $\mathcal{R}^{n \times n}$. Each $f_{\beta_n} \in L\infty([0, T], \mathcal{R}^n)$ and, since F is bounded, there is a constant C, such that $\|f_{\beta_n}(t)/C\| \le 1$. Thus, by Alaoglu's theorem, $\{f_{\beta_n}(t)/C\}$ is weakly* compact and there is a subsequence of $\{f_{\beta_n}\}$ weakly* converging to $h \in L^\infty([0], T, \mathcal{R}^n)$. Clearly, the mapping $T_\alpha : L^\infty([0], T, \mathcal{R}^n) \to C([0], T, \mathcal{R}^n) : g(\cdot) \mapsto \int_0^t \Phi_\alpha(t - s) \ g(s)ds$ is a compact operator, so the sequence $\{T_{\beta_n}, f_{\beta_n}\}$ is compact. Hence, there is a subsequence $\beta_{n(k)}$ such that $T_{\beta_{n(k)}} f_{\beta_{n(k))}} \to T_\beta h$, where $h(\cdot) \in F_\beta(\cdot)$. Along with the other convergences, clearly $x_{\beta_{n(k)}} \to \xi$ and $\xi \in S_\beta$ from (6.5.7). That is, $\tilde{S}_\beta(t) \subseteq S_\beta(t)$. By the Negoita–Ralescu characterization Theorem 1.5.1, the $S_\beta(t)$ are thus the level sets of a fuzzy set $S(t)$.

Example 6.5.2. *(Example 6.5.1 continued) For $0 \le t \le T < \infty$, let the nonhomogeneous term be given by*

$$F(t) = \left(\begin{array}{c} 1 + t; 2 + 3t)_s \\ (e^{-t}; e^t)_s \end{array} \right)$$

and consider the FDE

$$x'(t) = Ax(t) + F(t), \quad x(0) = X^0 = \begin{pmatrix} (1;2)_s \\ (0;1)_s \end{pmatrix},$$

where A is as in the first part of the example. Note that the level set transition matrices $\underline{\Phi}_\beta$, $\bar{\Phi}_\beta$ are positive matrices for $t > 0$ and when multiplying interval vectors, the order of end points is preserved. Writing $S_\beta(t) = [\underline{S}_\beta(t), \bar{S}_\beta(t)]$, from Theorem 6.5.2

$$\underline{S}_\beta(t) = \underline{\Phi}_\beta(t)\underline{X}_\beta^0 + \int_0^t \underline{\Phi}_\beta(t-s)\underline{F}_\beta(s)ds$$

$$\bar{S}_\beta(t) = \bar{\Phi}_\beta(t)\bar{X}_\beta^0 + \int_0^t \bar{\Phi}_\beta(t-s)\bar{F}_\beta(s)ds.$$

For $\beta = 0.0$, using the results in the first part of the example, a tedious calculation produces

$$\underline{S}_{0.0}(t) = \begin{pmatrix} -2.5308 - 1.1430t + 0.0826e^{-t} + 3.1007e^{\underline{\kappa}t} + 0.3476e^{\underline{\nu}t} \\ 0.2040 + 0.3734t + 0.3479e^{-t} - 0.5517e^{\underline{\kappa}t} - 0.0002e^{\underline{\nu}t} \end{pmatrix}$$

$$\bar{S}_{0.0}(t) = \begin{pmatrix} -3.9942 - 2.3797t - 6.0022e^t + 10.4599e^{\bar{\kappa}t} + 1.5363e^{\bar{\nu}t} \\ 2.3996 + 1.1998t + 4.0022e^t - 7.5399e^{\bar{\kappa}t} + 2.1380e^{\bar{\nu}t} \end{pmatrix},$$

where $\underline{\kappa} = 0.7929$, $\underline{\nu} = 2.2071$, $\bar{\kappa} = 0.9189$, $\bar{\nu} = 4.0811$. The same can be done at any β-level and, using interpolation, an estimation of $S(t)$ obtained. Closed form solutions for the β-levels can be obtained, but are quite complicated to write out.

6.6 Fuzzy Volterra Integral Equations

A number of papers [89, 95, 110] on fuzzy integral equations have recently appeared which interpret an integral equation as an equality involving the Aumann integral [2] at each level set. For example

$$y'(t) = G(t) + \int_0^t F(t, \sigma, y(\sigma))d\sigma, \qquad (6.6.1)$$

this representation meaning that

$$[y'(t)]^\beta = [G(t)]^\beta + \int_0^t [F(t, \sigma, y(\sigma))]^\beta d\sigma \quad 0 \le \beta \le 1.$$

Here, y is a fuzzy n-vector, G and F are fuzzy n-vector valued functions, $[\cdot]^\beta$ denotes the β-level set and the integral is that of Aumann.

Although impressive machinery from the crisp case can be adapted to (6.6.1) to prove existence, and sometimes uniqueness of solutions, the formulation involving the Aumann integral in this way suffers a fundamental flaw: The model simply does not reflect the behavior of real life systems with elements of uncertainty and vagueness, in very similar fashion to the corresponding failure of the classical formulation of fuzzy differential equations (FDEs). A simple example serves to illustrate the problem. Let $(c; d)_S$ denote the symmetric triangular fuzzy number with the interval $[c, d]$ as its support and let $x(t)$ be a fuzzy number valued function of time. Consider the integral equation

$$x(t) = (-1; 1)_S + \int_0^t (-x(\sigma)) d\sigma. \qquad (6.6.2)$$

Write the β-level set of $x(t)$ as the compact interval $x_\beta(t) = [\underline{x}_\beta(t), \bar{x}_\beta(t)]$ and note that $-x_\beta = [-\bar{x}_\beta, -\underline{x}_\beta]$. Using the Laplace transform and integral inequalities, see Lakshmikantham and Leela [61], one obtains the solution

$$x_\beta(t) = [(-1 + \beta)e^t, (1 - \beta)e^t], \ t \geq 0, \ 0 \leq \beta \leq 1.$$

Thus, (6.6.2) has an unstable solution, *in contrast to the behavior of the associated crisp integral equation* $z(t) = z_0 - \int_0^t (-z(\sigma)) d\sigma$, which has solution $z(t) = z_0 e^{-t}$. Simply by attributing some uncertainty to the initial value here, despite seemingly maintaining the same formal functional form of the equation, leads to behavior which is mathematically and intuitively quite different from the way the real world situation should be modeled.

If the integral equation (6.6.2) is replaced by the family of inclusions

$$y_\beta(t) \in [-1 + \beta, 1 - \beta] + \int_0^t (-y_\beta(\sigma)) d\sigma, \quad 0 \leq \beta \leq 1, \qquad (6.6.3)$$

and the solution of (6.6.3) is interpreted as the set $S_\beta(t)$ of all those $y_\beta(\cdot)$ which satisfy the β-th inclusion almost everywhere $\tau \in [0, t]$, then it has the fuzzy solution given by the β-level sets

$$S_\beta(t) = [-1 + \beta, 1 - \beta]e^{-t}, \quad 0 \leq \beta \leq 1,$$

which is asymptotically stable and reflects the behavior of the crisp equation despite uncertainty in the initial condition.

Let $C_n[0, T]$ be the space of continuous functions $f : [0, T] \to \mathcal{R}^n$, endowed with the norm of uniform convergence, $\|f - g\|_\infty = \max_{0 \leq t \leq T} |f(t) -$

$g(t)|$. Denote the space of normal, connected usc fuzzy sets, with compact support over $C_n[0, T]$, by $C^n[0, T]$ and give it the metric D_∞ induced by the Hausdorff metric on compact subsets of $C_n[0, T]$. The compact sets here are characterized by Arzela's theorem, as the families of continuous functions on $[0, T]$ which are *equicontinuous* and *uniformly bounded*.

Two further function spaces are needed.

(i) $L_n^1[0, T]$ is the space of *integrable functions* from $[0, T]$ to $\mathcal{R}^n$, with metric $\|f - g\|_1 = \int_0^T |f(t) - g(t)| dt$.

(ii) $L_n^\infty[0, T]$ is the space of measurable functions from $[0, T]$ to $\mathcal{R}^n$, bounded almost everywhere on $[0, T]$, with norm

$$\|f\|_{L^\infty} = \operatorname*{ess\,sup}_{0 \le t \le T} |f(t)| = \inf\{c \ge 0 : |f(t)| \le c \text{ a.e } 0 \le t \le T\}.$$

Here, the *essential supremum* is the greatest lower bound of all upper bounds taken except on a set of measure zero (the sets on which f is unbounded).

Note that the dual of $L_n^1[0, T]$ is $L_n^\infty[0, T]$, but that $L_n^1[0, T]$ is not reflexive. That is, $L_n^1[0, T]^* = L_n^\infty[0, T]$, but $L_n^1[0, T]^{**} \ne L_n^1[0, T]$. Also, $C_n[0, T]$ is a closed subspace of $L_n^\infty[0, T]$.

A sequence $\{x_n\}$ in a Banach space $\mathcal{X}$ is said to *converge weakly* to $x \in \mathcal{X}$ if $f(x_n) \to f(x)$ for every linear functional $f \in \mathcal{X}^*$. A sequence $\{f_n\}$ of linear functionals in $\mathcal{X}^*$ converges weakly* to $f \in \mathcal{X}$ if $f_n(x) \to f(x)$ for every $x \in \mathcal{X}$. Note that the unit ball of $\mathcal{X}^*$ is compact in the weak* topology, a result called Alaoglu's theorem. In particular, the unit ball in $L_n^\infty[0, T]$ is weak* compact.

Remark 6.6.1. We note that the conclusions of the stacking theorem 6.2.2 also hold if the U_β are subsets of $C_n[0, T]$, satisfying the same conditions, and then $X \in C^n[0, T]$. The converse holds as well.

The literature on integral inclusions is much sparser than that for integral equations, despite there being significant applications. The treatment we follow here is adapted from Corduneanu [12]. The inclusions considered take the form, first

$$x(t) \in h(t) + \int_0^t k(t, s)G(s, x(s))ds \quad t \in [0, T], \tag{6.6.4}$$

where $h : [0, T] \to \mathcal{R}^n$ is continuous, the matrix valued function $k : \{(s, t : 0 \le s \le t \le T\} \to L_{n \times n}^1[0, T]$, $G : [0, T] \times \mathcal{R}^n \to \mathcal{K}_c^n$ and $G(\cdot, x)$ has measurable selections. The integral is understood to be in the sense of Aumann. That is, if $F : [0,T] \to \mathcal{K}_C^n$ and $\mathcal{S}(F)$ denotes the set of integrable selections of F, then

$$\int_0^t F(s)ds = \left(\int_0^t f(s)ds : f \in \mathcal{S}(F) \right).$$

A solution to (6.6.4) is a continuous function $x(\cdot) \in C[0,T]$ which satisfies the inclusion a.e. in $[0,T]$. Denote the set of solutions in $[0,\tau]$ by $S(h;\tau)$ and the attainability set for each $\tau \in [0,T]$ by $\mathcal{A}(h;\tau) = \{x(\tau) : x(\cdot) \in S(h;\tau)\}$. It is known (see Corduneanu [12]) that, under mild further conditions on k and G, solutions exist and the sets $S(h;\tau)$, $\mathcal{A}(h;\tau)$ are not only nonempty, but also compact and connected. If $M \subset C_n[0,T]$ and $0 \le \tau \le T$, write $S(M,\tau) = \bigcup_{h \in M} S(h;\tau)$ and $\mathcal{A}(M;\tau) = \bigcup_{h \in M} \mathcal{A}(h;\tau)$.

Theorem 6.6.1. *Let* $G : [0,T] \times \mathcal{R}^n \to \mathcal{K}_C^n$ *and suppose that the following conditions hold:*

   (i) $G(\cdot,x)$ *is measurable, for every* $x \in \mathcal{R}^n$.

  (ii) $G(t,\cdot)$ *is usc for a.e.* $t \in [0,T]$.

 (iii) *There exists* $p \in L^1[0,T]$ *with* $\|G(t,x)\| \le |p(t)|$ *for a.e.* $t \in [0,T]$ *and every* $x \in \mathcal{R}^n$.

  (iv) $h \in C_n[0,T]$.

   (v) *For each* $t \in [0,T]$, $k(t,s)$ *is a measurable function of* $s \in [0,T]$ *and* $k(t) = \text{ess}\sup_{0 \le s \le t} |k(t,s)|$ *is bounded on* $[0,T]$.

  (vi) *The map* $t \mapsto k_t$ *is continuous from* $[0,T]$ *to* $L_n^\infty[0,T]$, *where* $k_t(s) = k(t,s)$.

Then for every $h \in C_n[0,T]$, *the set* $S(h;\tau)$, $0 \le \tau \le T$, *is nonempty, connected, and compact. For every compact, connected subset* $M \subset C_n[0,T]$, *the set* $S(M;T)$ *is compact and connected. The map* $h \mapsto S(h;T)$ *is usc. The same conclusions hold for* $\mathcal{A}(h;\tau)$ *and* $\mathcal{A}(M;\tau)$, $0 \le \tau \le T$.

For a proof, see Corduneanu [12].

Remark 6.6.2. If $M \subset C_n[0,T]$, the inclusion

$$x(\cdot) \in M + \int_0^{\cdot} k(t,s)G(s,x(s))ds$$

will be understood to mean that there exists $h \in M$ such that $x(t) \in h(t) +$ $\int_0^t k(t,s)G(s,x(s))ds$ a.e. $t \in [0,T]$.

The following lemma is also from Corduneanu [12] which is useful in the proof of Theorem 6.6.2.

Lemma 6.6.1. *Define the linear Volterra operator W by*

$$(Wh)(t) = \int_0^t k(t,s)h(s)ds.$$

Then, under assumptions (v) and (vi) of Theorem 6.6.1, if $h_k \rightarrow h$ weakly in $L_n^1[0,T]$ as $k \rightarrow \infty$, then there exists a subsequence $\{h_{k(m)}\} \subset \{h_k\}$ such that $Wh_{k(m)} \rightarrow Wh$ in $C_n[0,T]$ as $k(m) \rightarrow \infty$.

Now we are in a position to investigate fuzzy integral equations. Let $V \in C^n[0,T]$, that is, V is a connected usc fuzzy set with compact support over the space of continuous functions $C_n[0,T]$. Then the β-level sets are equibounded and equicontinuous connected sets of continuous functions. Suppose that $V(t)$ is the value at t and $F : [0,T] \times \mathcal{R}^n \rightarrow E^n$ and consider the fuzzy integral equation

$$x(t) = V(t) + \int_0^t k(t,s)F(s,x(s))ds. \qquad (6.6.5)$$

As was observed earlier, this formulation has significant problems. Instead, interpret the equation (6.6.5) as a family of integral inclusions

$$x_\beta(\cdot) \in V_\beta + \int_0^{\cdot} k(t,s)F_\beta(s,x(s))ds, \quad 0 \le \beta \le 1, \qquad (6.6.6)$$

where the subscript β indicates that the β-level set of a fuzzy set is involved. Denote the solution set of the β-th inclusion by $S_\beta(V_\beta;t)$, $0 \le t \le T$, and the corresponding attainability set by $\mathcal{A}_\beta(V_\beta;t)$. The system (6.6.6) can only have any significance as a replacement for (6.6.5) if the solutions generate fuzzy sets. That is, if the sets S_β, $\mathcal{A}_\beta$ are level sets of fuzzy sets S and $\mathcal{A}$, respectively. The existence of compact connected S_β and $\mathcal{A}_\beta$ follow from Theorem 6.6.1. It remains to show that these sets satisfy the conditions of the stacking theorem 6.2.2 over $C_n[0,T]$ and $\mathcal{R}^n$, respectively. The first condition, that they are nonempty, compact and connected, follows from Theorem 5.6.1 while the second condition $S_\beta \subseteq S_\alpha$ for all $0 \le \alpha \le \beta \le 1$ easily follows because $V_\beta \subseteq V_\alpha$ and $F_\beta \subseteq F_\alpha$. The only part that requires some demonstration is the third condition. One further concept is required: a mapping $H : [0,T] \rightarrow E^n$ is *strongly measurable* if each level set mapping $[H(\cdot)]^{\beta} : [0, T] \to \mathcal{K}_C^n$ is measurable (see Diamond and Kloeden [24], Example 10.1.3, for an example of a measurable mapping which is not strongly measurable). The following theorem gives the fuzzy analogs of the assumptions of Theorem 5.6.1. Below, $\tilde{0}$ is the fuzzy set defined by $\tilde{0}(\xi) = 1$ if $\xi = 0$ and $\tilde{0}(\xi) = 0$ if $\xi \neq 0$.

Theorem 6.6.2. *Let $F : [0, T] \times \mathcal{R}^n \to E^n$ and suppose that the following conditions hold for the system of inclusions (6.6.6):*

(i)′ *$F(\cdot, x)$ is strongly measurable for every $x \in \mathcal{R}^n$.*

(ii)′ *$F(t, \cdot)$ is usc for a.e. $t \in [0, T]$.*

(iii)′ *There exists $p \in L^1[0, T]$ with $\|F(t, x)\| \leq |p(t)|$ for a.e. $t \in [0, T]$ and every $x \in \mathcal{R}^n$, where $\|F\| = d(F, \tilde{0})$.*

(iv)′ *$V \in \mathcal{C}^n[0, T]$.*

(v)′ *For each $t \in [0, T]$, $k(t, s)$ is a measurable function of $s \in [0, t]$ and $k(t) = \text{ess sup}_{0 \leq s \leq t} |k(t, s)|$ is bounded on $[0, T]$, $|k(t)| \leq Q$.*

(vi)′ *The map $t \mapsto k_t$ is continuous from $[0, T]$ to $L_n^\infty[0, T]$, where $k_t(s) = k(t, s)$.*

Then the solution sets $S_\beta(V_\beta; t)$, $0 \leq t \leq T$, of the family of inclusions (6.6.6) are the level sets of a fuzzy set $S(V; t) \in \mathcal{C}^n[0, t]$ and the attainability sets $\mathcal{A}_\beta(V_\beta; t)$ are the level sets of a fuzzy set $\mathcal{A}(V; t) \in \mathcal{D}^n$.

Proof.   First, the sets $S(V_\beta; t)$ and $\mathcal{A}_\beta(V_\beta; t)$ are nonempty, compact and connected. This follows because condition (i) of Theorem 6.6.1 is satisfied by (i)′, condition (ii) of that theorem follows from (ii)′ and Lemma 6.2.1, and (iii) is implied by (iii)′, since $\|F_\beta\| \leq \|F\|$. So the first condition of the stacking theorem 6.2.2 holds. Second, since $V_\beta \subseteq V_\alpha$ and $G_\beta \subseteq G_\alpha$ for $0 \leq \alpha \leq \beta \leq 1$, obviously also $S_\beta(V_\beta; t) \subseteq S_\alpha(V_\alpha; t)$ and similarly for the $\mathcal{A}_\beta$ sets, which is the second condition. It remains to show that if $\beta_i \to \beta$ is any nondecreasing sequence in $[0, 1]$, then

$$S_\beta(V_\beta; t) = \bigcap_{i=1}^{\infty} S_{\beta_i}(V_{\beta_i}; t).$$

Now, $S_{\beta_i}(V_{\beta_i}; t)$ is a nonincreasing sequence of nonempty compact, connected sets and so $\bigcap_i S_{\beta_i}(V_{\beta_i}; t) = \tilde{S}$ is nonempty, connected and compact. Furthermore, $d_H^*(S_{\beta_i}(V_{\beta_i}; t), \tilde{S}) \to 0$ as $i \to \infty$, where d_H^* is the Hausdorff metric on compact subsets of $C_n[0,T]$. To see that $S_\beta(V_\beta;t) = \tilde{S}$, it suffices to show $S_\beta(V_\beta;t) \supseteq \tilde{S}$ since $S_\beta(V_\beta;t) \subset \tilde{S}$ is clear. For each i, let $x_{\beta_i} \in S_{\beta_i}(V_{\beta_i};t)$. The inclusion (6.6.6) means that there is a continuous function $v_{\beta_i} \in V_{\beta_i}$ and an integrable function $u_{\beta_i}(\cdot) \in F_{\beta_i}(\cdot, x_{\beta_i}(\cdot))$ such that

$$x_{\beta_i}(t) = v_{\beta_i}(t) + \int_0^t k(t,s) u_{\beta_i}(s) ds. \tag{6.6.7}$$

Since the support of V, $V_0 \supseteq V_{\beta_i}$ is compact, by Arzela's theorem the sequence $\{v_{\beta_i}\}$ is uniformly bounded and equicontinuous. So, from (iii)$'$ and (v)$'$,

$$
\begin{aligned}
|x_{\beta_i}(t)| &\leq |v_{\beta_i}(t)| + \left(\sup_{0 \leq t \leq T} k(t) \right) \int_0^T |u_{\beta_i}(s)| ds \\
&\leq |v_{\beta_i}| + Q \int_0^T p(s) ds
\end{aligned}
$$

and so $\{x_{\beta_i}\}$ is uniformly bounded. Again,

$$
\begin{aligned}
|x_{\beta_i}(t) - x_{\beta_i}(\tau)| &\leq |v_{\beta_i}(t) - v_{\beta_i}(\tau)| + \int_0^t |k(t,s) - k(\tau,s)||u_{\beta_i}(s)| ds \\
&\quad + \int_\tau^t |k(\tau,s)||u_{\beta_i}(s)| ds \\
&\leq |v_{\beta_i}(t) - v_{\beta_i}(\tau)| + \|k(t,\cdot) - k(\tau,\cdot)\|_{L^\infty} \int_0^T p(s) ds \\
&\quad + Q \int_\tau^t p(s) ds.
\end{aligned}
$$

From (vi)$'$, k_t is uniformly continuous because $[0,T]$ is compact. Since $\{v_{\beta_i}\}$ is equi-continuous, it follows that $\{x_{\beta_i}\}$ is equi-continuous and so compact. Hence, there exists a subsequence $\{x_{\beta_{i(1)}}\} \subset \{x_{\beta_i}\}$ such that $x_{\beta_{i(1)}} \to x_\beta \in C_n[0,T]$. From the compactness of $\{v_{\beta_i}\}$, $\{v_{\beta_{i(1)}}\}$ is also compact, so there exists a further subsequence $v_{\beta_{i(2)}} \to v_\beta \in V_\beta$ and clearly $v_{\beta_{i(2)}} \to x_\beta$.

Since $\|u_{\beta_{i(2)}}(t)\| \leq p(t)$, the sequence of functions $w_{\beta_{i(2)}}(t) = u_{\beta_{i(2)}}(t)/p(t)$ belongs to the unit ball of $L_n^\infty[0,T]$, which is weakly* compact by Alaoglu's theorem, so a subsequence $\{w_{\beta_{i(3)}}\}$ converges weakly* to $w_\beta \in L_n^\infty[0,T]$. But the map $w \mapsto p(t)w$ is a continuous map from L_n^∞ to L_n^1 and so the sequence $\{u_{\beta_{i(3)}}\}$ converges weakly in $L_n^1[0,T]$ to $u_\beta = pw_\beta$.

Now, from conditions (v)$'$, (vi)$'$ and Lemma 6.6.1, there exists yet a further subsequence $\{u_{\beta_k}\} \subset \{u_{\beta_{i(3)}}\}$ such that

$$\int_0^t k(t,s)u_{\beta_k}(s)ds \to \int_0^t k(t,s)u_\beta(s)ds.$$

Hence,

$$x_\beta(t) = v_\beta(t) + \int_0^t k(t,s)u_\beta(s)ds,$$

where $u_\beta \in F_\beta(\cdot, x_\beta(\cdot))$ by the usc of F. That is, (6.6.6) is satisfied and so $x_\beta \in S_\beta(V_\beta; t)$. That is $\tilde{S} \subseteq S_\beta(V_\beta; t)$ and the theorem is proved.

Before providing the examples, recall (Lakshmikantham and Leela [61], pp. 315–318) that if $K(t,s,x)$ is nondecreasing in x and

$$x(t) = h(t) + \int_0^t K(t,s,x(s))ds$$

$$y(t) \geq h(t) + \int_0^t K(t,s,y(s))ds,$$

then $x(t) \leq y(t)$, $t \geq 0$. Conversely, if K is nonincreasing in x and the inequality in the second equation is reversed, then $x(t) \geq y(t)$ for $t \geq 0$.

Example 6.6.1. *Consider the fuzzy integral equation*

$$x(t) = (0; t^2/2)_S - \int_0^t (t-\sigma)(x(\sigma); 3x(\sigma))_S d\sigma, \quad t \geq 0,$$

or, equivalently from the preceding sections, for $0 \leq \beta \leq 1$,

$$x_\beta(t) \in [\beta t^2/4, (1/2 - \beta/4)t^2] - \int_0^t (t-\sigma)[(1+\beta)x_\beta(\sigma), (3-\beta)x_\beta(\sigma)]d\sigma.$$

That is,

$$x_\beta(t) \geq \beta t^2/4 - (3-\beta)\int_0^t (t-\sigma)x_\beta(\sigma)d\sigma$$

$$x_\beta(t) \leq (1/2 - \beta/4)t^2 - (1+\beta)\int_0^t (t-\sigma)x_\beta(\sigma)d\sigma.$$

The inequality makes sense, since both $1+\beta \leq 3-\beta$ and $\beta/4 \leq 1/2 - \beta/4$ for $0 \leq \beta \leq 1$.

Taking Laplace transforms, noting the convolution integrals and using the result quoted on integral inequalities

$$\frac{\beta}{2s^3} - \frac{(3-\beta)X_\beta(s)}{s^2} \leq X_\beta(s) \leq \frac{1-\beta 2}{s^3} - \frac{(1+\beta)X_\beta(s)}{s^2}.$$

Clearly,

$$\frac{\beta/2}{s(s^2 + 3 - \beta)} \leq X_\beta(s) \leq \frac{1 - \beta/2}{s(s^2 + 1 + \beta)}. \qquad (6.6.8)$$

Note that

$$\frac{a}{s(s^2 + b)} = \frac{a}{bs} - \frac{as/b}{s^2 + b},$$

apply it to (6.6.8) and take the inverse transform to obtain

$$\frac{\beta(1 - \cos(\sqrt{3 - \beta}t))}{2(3 - \beta)} \leq x_\beta(t) \leq \frac{(2 - \beta)(1 - \cos(\sqrt{1 + \beta}t))}{2(1 + \beta)}.$$

That is, the solution set $S(t)$ consists of the fuzzy set with β-levels the intervals

$$\left[\frac{\beta(1 - \cos(\sqrt{3 - \beta}t))}{2(3 - \beta)}, \frac{(2 - \beta)(1 - \cos(\sqrt{1 + \beta}t))}{2(1 + \beta)}, \right]$$

for $0 \leq \beta \leq 1$.

Example 6.6.2. *Recall that an n-dimensional open loop linear control system*

$$x'(t) = Ax(t) + Bu(t), \quad x(0) = x_0,$$

has a solution which can be written in terms of the variation of constants formula as

$$x(t) = \Phi(t)x_0 + \int_0^t \Phi(t - s)Bu(s)ds. \qquad (6.6.9)$$

Here, $\Phi(t)$ is the state transition matrix, or matrix exponential, satisfying the matrix differential equation

$$\Phi'(t) = A\Phi(t), \quad \Phi(0) = I. \qquad (6.6.10)$$

   Now, suppose that the matrices A, B have fuzzy number entries and the initial condition $x_0 \in E^n$ is fuzzy. Then (6.6.9) can be considered as a family of integral inclusions, provided that some meaning can be ascribed to $\Phi(t)$. Following the earlier discussion, we interpret (6.6.10) as the family of differential inclusions

$$\Phi'_\beta(t) \in A_\beta \Phi_\beta(t), \quad \Phi_\beta(0) = I, \quad 0 \leq \beta \leq 1, \qquad (6.6.11)$$

where Φ_β, A_β denote level sets. That is, if V is a matrix in the set of matrices A_β,

$$\Phi_\beta(t) = \{Y(t) : Y' = VY, Y(0) = I\}.$$

The $Y(t)$ are found in the usual way: find a basis of vector solutions $v_1(t)$,
$v_2(t)$, ..., $v_n(t)$ from the eigenvalue–eigenvector problem for V and form the
matrix $Z(t) = [v_1 \ v_2 \ ... \ v_n]$. Then $Y(t) = Z(t)Z(0)^{-1}$.

Since A_β is an interval matrix, that is, has compact real intervals as
entries, U belongs to the interval $[\underline{A}_\beta, \bar{A}_\beta]$. Here, $\underline{A}$, $\bar{A}$ denote ordinary
matrices whose elements are, respectively, the lower and upper end points of
the real intervals (see Neumaier [80] for notation and theory). This gives
a method for evaluating $\Phi_\beta(\cdot)$ as an interval matrix. by a simple extension
of the results in Sections 5.2 and 5.6, the $\Phi_\beta(t)$ form the level sets of a
$\mathcal{D}^{n\times n}$-valued function.

In the case where $\underline{A}_\beta$ is a nonnegative matrix, that is, all elements of the
matrix are nonnegative, or $\bar{A}_\beta$ is a nonpositive matrix, the computation is
especially simple. If $C - A$ is a nonnegative matrix, write $C \geq A$.

Lemma 6.6.2. *Let A be a nonnegative matrix and suppose that $C \geq A$. If*

$$X'(t) = AX, \quad X(0) = I,$$

$$Y;(t) = CY, \quad Y(0) = I,$$

then $Y(t) \geq X(t)$, $t \geq 0$. In particular, if $A = [\underline{A}, \bar{A}]$ is an interval matrix
with $0 \leq \underline{A} \leq V \leq \bar{A}$ and

$$\underline{X}'(t) = \underline{A}\underline{X}(t), \quad X'(t) = VX(t), \quad \bar{X}'(t) = \bar{A}\bar{X}(t),$$

$$\underline{X}(0) = X(0) = \bar{X}(0) = I$$

then $\underline{X}(t) \leq X(t) \leq \bar{X}(t)$, $t \geq 0$.

Proof.   This is a simple consequence of the result mentioned earlier for
integral inequalities, since the matrix differential equations are equivalent to

$$X(t) = I + \int_0^t AX(s)ds,$$

$$Y(t) = I + \int_0^t CY(s)ds \geq I + \int_0^t AY(s)ds,$$

and the function $X \mapsto AX$ is monotonic nondecreasing in the partial order
induced by the positive orthant.

Clearly, a similar result holds for nonpositive matrices. In the case where
the interval matrix is not of these types, the interval matrix function will, in
general, have end points corresponding to matrices internal to the internal

matrix A and can be estimated numerically by solving the matrix DEs on a grid.

As a numerical illustration, consider the system where $n = 2$ and A, B, $t \geq 0$, x_0 are given by

$$A = \begin{pmatrix} (-2.5; -1.5)_S & (-1.4; -0.6)_S \\ (-1.4; -0.6)_S & (-3.7; -2.3)_S \end{pmatrix},$$

$$B = \begin{pmatrix} (2; 3)_S \\ (1; 2)_S \end{pmatrix},$$

$$x_0 = \begin{pmatrix} (1; 2)_S \\ (0; 1)_S \end{pmatrix},$$

with a scalar control law given by the fuzzy-valued function

$$u(t) = (e^{t/2}; 2e^t)_S.$$

Expressing the β-levels of A as an interval matrix:

$$A_\beta = \begin{pmatrix} (-2.5 + 0.5\beta, -1.5 - 0.5\beta) & (-1.4 + 0.4\beta, -0.6 - 0.4\beta) \\ (-1.4 + 0.4\beta, -0.6 - 0.4\beta) & (-3.7 + 0.7\beta, -2.3 - 0.7\beta) \end{pmatrix},$$

this interval system matrix is a stable family, because every characteristic polynomial is quadratic with positive coefficients. So, for example, if $\beta = 0.0$, $\Phi_{0.0} = [\underline{\Phi}_{0.0}, \bar{\Phi}_{0.0}]$. Applying Lemma 6.6.2 using the MATLAB functions EIG and INV for the eigenvalue–eigenvector and matrix inversion calculations

$$\underline{\Phi}_{0.0} = \begin{pmatrix} 0.6969e^{\underline{\kappa}t} + 0.3031e^{\underline{\nu}t} & -0.4596e^{\underline{\kappa}t} + 0.4596e^{\underline{\nu}t} \\ -0.4596e^{\underline{\kappa}t} + 0.4596e^{\underline{\nu}t} & 0.3031e^{\underline{\kappa}t} + 0.6969e^{\underline{\nu}t} \end{pmatrix}, \quad (6.6.12)$$

where $\underline{\kappa} = -1.5768$, $\underline{\nu} = -4.6232$, and

$$\bar{\Phi}_{0.0} = \begin{pmatrix} 0.7774e^{\bar{\kappa}t} + 0.2227e^{\bar{\nu}t} & -0.4161e^{\bar{\kappa}t} + 0.4161e^{\bar{\nu}t} \\ -0.4161e^{\bar{\kappa}t} + 0.4161e^{\bar{\nu}t} & 0.2227e^{\bar{\kappa}t} + 0.7774e^{\bar{\nu}t} \end{pmatrix}, \quad (6.6.13)$$

where $\bar{\kappa} = -1.1789$, $\bar{\nu} = -2.6211$. (INV was not really needed because the end point matrices of the interval are symmetric.)

Now, turning to the formula (6.6.9), with both the control system equation and Φ interpreted in the differential inclusion sense (6.6.11), and writing $x_\beta(t) = [\underline{x}_\beta(t), \bar{x}_\beta(t)]$, the following is obtained

$$x_\beta(t) = [\underline{\Phi}_\beta(t), \bar{\Phi}_\beta(t)][x_0]^\beta + \int_0^t [\underline{\Phi}_\beta(t - \sigma), \bar{\Phi}_\beta(t - \sigma)][\underline{B}_\beta, \bar{B}_\beta]u_\beta(\sigma)d\sigma,$$

where

$$u_\beta(\sigma) = [\underline{u}_\beta, \bar{u}_\beta(\sigma)]$$
$$= [e^{\sigma/2} + \beta(2e^\sigma - e^{\sigma/2})/2, 2e^\sigma - \beta(2e^\sigma - e^{\sigma/2})/2].$$

Since the integral is a convolution, when the Laplace transform is taken

$$X_\beta(s) = \phi_\beta(s)[\underline{x}_{0\beta}, \bar{x}_{0\beta}] + \phi_\beta(s)BU_\beta(s). \qquad (6.6.14)$$

Here, $\phi_\beta(s) = [\underline{\phi}_\beta(s), \bar{\phi}_\beta(s)]$, $U_\beta(s) = [\underline{U}_\beta(s), \bar{U}_\beta(s)]$ are, respectively, the transforms of $\Phi_\beta(t)$, $u_\beta(t)$. A straightforward, but tedious, calculation from (6.6.14) when $\beta = 0.0$ gives

$$\underline{x}_{0.0}(t) = \begin{pmatrix} 0.2471e^{\underline{\kappa}t} + 0.0951e^{\underline{\nu}t} + 0.6578e^{t/2} \\ -0.1629e^{\underline{\kappa}t} + 0.1442e^{\underline{\nu}t} + 0.0817e^{t/2} \end{pmatrix}$$

$$\bar{x}_{0.0}(t) = \begin{pmatrix} -0.2381e^{\bar{\kappa}t} - 0.2833e^{\bar{\nu}t} + 2.5216e^t \\ 0.8144e^{\bar{\kappa}t} - 0.5293e^{\bar{\nu}t} + 0.7150e^t \end{pmatrix}.$$

6.7   Notes and Comments

As we have seen, Hüllermeir [40] suggested a different formulation of the fuzzy initial value problem based on a family of differential inclusions at each β-level, $0 \le \beta \le 1$, namely,

$$x'(t) \in [G(t, x(t)]^\beta, \quad x(0) = [x_0]^\beta,$$

where $[G(\cdot, \cdot)]^\beta : \mathcal{R} \times \mathcal{R}^n \to K_C^n$. However, Hüllermeir does not prove that $S(x_0, T)$ and $\mathcal{A}(x_0, t)$ are fuzzy sets and moreover requires that $[G(t, x)]^\beta$ be not only bounded but also continuous and Lipschitz in x with respect to d_H. The results presented in Section 6.2 are taken from Diamond and Watson [25] where to formulate the problem of fuzzy differential inclusions, the notion of quasi-concavity is employed to obtain regularity of solution sets. See also Diamond [18]. Section 6.3 essentially lists the needed results for differential inclusions from Deimling [17]. In Section 6.4, fuzzy differential inclusions are discussed and periodicity and stability results are presented, which are taken from Diamond [19]. The variation of constants formula considered in Section 6.5 is from Diamond [21]. See Rzezuchowski and Wasowski [103] for the results on continuous dependence and parameters and initial values of solutions of differential equations with fuzzy parameters via differential inclusions. Finally, for the results related to fuzzy Volterra integral inclusions given in Section 6.6, see Diamond [20] which depends on the corresponding results on integral equations that are taken from Corduneanu [12]. For allied results, see also Diamond [22].

Bibliography

[1] Aubin, J.P. and Cellina, A., *Differential Inclusions*, Springer Verlag, New York 1984.

[2] Aumann, R.J., Integrals of set-valued functions, *J. Math. Anal. Appl.* **12** (1965), 1–12.

[3] Banks, H.T. and Jacobs, M.Q., A differential calculus of multifunctions, *JMAA* **29** (1970), 246–272.

[4] Bernfeld, S. and Lakshmikantham, V., *An Introduction to Nonlinear Boundary Value Problems*, Academic Press, New York 1974.

[5] Bobylev, V.N., Cauchy problem under fuzzy control, *BUSEFAL* **21** (1985), 117–126.

[6] Bobylev, V.N., A possibilistic argument for irreversibility, *Fuzzy Sets and Systems* **34** (1990), 73–80.

[7] Bradley, M. and Datko, R., Some analytic and measure theoretic properties of set-valued mappings, *SIAM J. Control. Optim.* **15** (1977), 625–635.

[8] Buckley, J.J. and Feuring, T., Fuzzy differential equations, *Fuzzy Sets and Systems* **110** (2000), no. 1, 43–54.

[9] Castaing, C. and Valadier, M., *Convex Analysis and Measurable Multifunctions*, Springer-Verlag, Berlin 1977.

[10] Chen, M., Saito, S., and Ishii, H., Representation of fuzzy numbers and fuzzy differential equations, *Mathematical Science of Optimization* (Japanese) (Kyoto, 2000).

[11] Constantin, A., Stability of solution sets of differential equations with multi-valued right-hand side, *J. Diff. Eq.*, **114** (1994), 243–252.

[12] Corduneanu, C., *Integral Equations and Applications*, Cambridge University Press, New York 1991.

[13] Corduneanu, C., *Functional Equations with Causal Operators*, Gordon and Breach, to appear.

[14] De Blasi, F.S. and Lasota, A., Daniell's method in the theory of the Aumann–Hukuhara integral of set-valued functions, *Atti Acad. Naz. Lincei Rendiconti Ser.* 8 **45** (1968), 252–256.

[15] De Blasi, F.S. and Myjak, J., On the solution sets for differential inclusions, *Bull. Acad. Polon. Sci.*, **33** (1985), 17–23.

[16] Debreu, G., Integration of correspondences, in: Proc. Fifth Berkeley Symp. Math. Statist. Probab., Vol. 2, Part #1, Univ. California Press, Berkeley, CA (1967), 351–372.

[17] Deimling, K., *Multivalued Differential Equations*, Walter de Gruyter, New York 1992.

[18] Diamond, P., Time-dependent differential inclusions, cocycle attractors and fuzzy differential equations, *IEEE Trans. Fuzzy Systems*, **7** (1999), 734–740.

[19] Diamond, P., Stability and periodicity in fuzzy differential equations, *IEEE Trans. Fuzzy Systems* **8** (2000), 583–590.

[20] Diamond, P., Theory and application of fuzzy Volterra integral equations (to appear).

[21] Diamond, P., Brief note on the variation of constants formula for fuzzy differential equations (to appear).

[22] Diamond, P., Design of optimally bounded linear state feedback laws for fuzzy dynamic systems (to appear).

[23] Diamond, P. and Kloeden, P., Characterization of compact subsets of fuzzy sets, *Fuzzy Sets and Systems* **29** (1989), 341–348.

[24] Diamond, P. and Kloeden, P., *Metric Spaces of Fuzzy Sets*, World Scientific, Singapore 1994.

[25] Diamond, P. and Watson, P., Regularity of solution sets for differential inclusions quasi-concave in parameter, *Appl. Math. Lett.*, **13** (2000), 31–35.

[26] Ding, Z. and Kandel, A., Existence and stability of fuzzy differential equations, *J. Fuzzy Math.* **5** (1997), no. 3, 681–697.

[27] Ding, Z., Ma, M., and Kandel, A., Existence of the solutions of fuzzy differential equations with parameters, *Inform. Sci.* **99** (1997), no. 3-4, 205–217.

[28] Driankov, D., Hellendoorn, H. and Reinfrank, M., *An Introduction to Fuzzy Control*, Springer-Verlag, Berlin 1996 (Second Edition).

[29] Dubois, D. and Prade, H., Fuzzy Sets and Systems: Theory and Applications, Volume 144 of *Mathematics in Science and Engineering*, Academic Press, New York (1980).

[30] Dubois, D. and Prade, H., Towards fuzzy differential calculus, Part I, *Fuzzy Sets and Systems* **8** (1982), 1–17.

[31] Dubois, D. and Prade, H., Towards fuzzy differential calculus, Part II, *Fuzzy Sets and Systems* **8** (1982), 105–116.

[32] Dubois, D. and Prade, H., Towards fuzzy differential calculus, Part III, *Fuzzy Sets and Systems* **8** (1982), 225–234,

[33] Friedman, M., Ma, M., and Kandel, A., Comments on: "The Peano theorem for fuzzy differential equations revisited", *Fuzzy Sets and Systems* **98** (1998), 147–148.

[34] Friedman, M., Ma, M. and Kandel, A., On the validity of the Peano theorem for fuzzy differential equations, *Fuzzy Sets and Systems* **86** (1997), no. 3, 331–334.

[35] Goetschel, R. and Voxman, W., A pseudometric for fuzzy sets and certain related results, *J. Math. Anal. Appl.* **81** (1981), 507–523.

[36] Goetschel, R. and Voxman, W., Topological properties of fuzzy numbers, *Fuzzy Sets and Systems* **10** (1983), 87–99.

[37] Hausdorff, F., *Set Theory*, Chelsea, New York 1957.

[38] Heilpern, S., Fuzzy mappings and fixed point theorem, *JMAA* **83** (1981), 566–569.

[39] Hukuhara, M., Integration des applications measurables dont la valeur est. un compact convexe, *Funkcialaj, Ekvacioj* **10** (1967), 205–223.

[40] Hüllermeier, E., An approach to modelling and simulation of uncertain dynamical systems, *Int. J. Uncertainty, Fuzziness & Knowledge-Based Systems*, **5** (1997), 117–137.

[41] Kaleva, O., On the convergence of fuzzy sets, *Fuzzy Sets and Systems* **17** (1985), 53–65.

[42] Kaleva, O., Fuzzy differential equations, *Fuzzy Sets and Systems* **24** (1987), no. 3, 301–317.

[43] Kaleva, O., The Cauchy problem for fuzzy differential equations, *Fuzzy Sets and Systems* **35** (1990), no. 3, 389–396.

[44] Kaleva, O., The Peano theorem for fuzzy differential equations revisited, *Fuzzy Sets and Systems* **98** (1998), no. 1, 147–148.

[45] Kaleva, O. and Seikkala, S., On fuzzy metric spaces, *Fuzzy Sets and Systems* **12** (1984), 215–229.

[46] Kandel, A., *Fuzzy Mathematical Techniques with Applications*, Addison Wesley, New York 1986.

[47] Kandel, A. and Byatt, W.J., Fuzzy differential equations, *Proceedings of the International Conference on Cybernetics and Society, Vol. I, II* (Tokyo/Kyoto, 1978), 1213–1216, *IEEE*, New York 1978.

[48] Kandel, A., Friedman, M., and Ma, M., On fuzziness and duality in fuzzy differential equations, Applications of fuzzy theory to complex systems (Taipei, 1995), *Internat. J. Uncertain. Fuzziness Knowledge-Based Systems* **4** (1996), no. 6, 553–560.

[49] Kauffman, A., *Introduction to the Theory of Fuzzy Subsets*, Vol. I, Academic Press, New York 1975.

[50] Kauffman, A. and Gupta, M.M., *Fuzzy Mathematical Models in Engineering and Management Sciences*, North-Holland, New York 1988.

[51] Klir, G.J. and Yuan, B., *Fuzzy Sets and Fuzzy Logic*, Prentice Hall, New York 1995.

[52] Kloeden, P.E., Compact supported endographs and fuzzy sets, *Fuzzy Sets and Systems*, **4** (1980), 193–201.

[53] Kloeden, P.E., Remarks on Peano-like theorems for fuzzy differential equations, *Fuzzy Sets and Systems* **44** (1991), no. 1, 161–163.

[54] Kruse, R., Gebhardt, J. and Klawonn, F., *Foundations of Fuzzy Systems*, John Wiley, New York 1994.

[55] Kwun, Y.-C., Kang, J.-R. and Kim, S.-H., Existence and uniqueness of fuzzy solution for nonlinear fuzzy differential equations, *Far East J. Math. Sci. (FJMS)* **1** (1999), no. 3, 487–500.

[56] Kwun, Y.-C. and Jeong, J.M., Control problem for fuzzy differential equations, *Differential Equations and Applications* (Chinju, 1998), 193–203, *Nova Sci. Publ.*, Huntington, New York 2000.

[57] Lakshmikantham, V., Uncertain systems and fuzzy differential equations, *J. math. Anal. Appl.* **251** (2000), no. 2, 805–817.

[58] Lakshmikantham, V., Bainov, D.D. and Simeonov, P.S., *Theory of Impulsive Differential Equations*, World Scientific, Singapore 1989.

[59] Lakshmikantham, V. and Leela, S., *Stability Analysis of Nonlinear Systems*, Marcel Dekker, 1969.

[60] Lakshmikantham, V. and Leela, S., *Nonlinear Differential Equations in Abstract Spaces*, Pergamon Press, Oxford 1981.

[61] Lakshmikantham, V. and Leela, S., *Differential and Integral Inequalities*, Vol. I, Academic Press, New York 1996.

[62] Lakshmikantham, V. and Leela, S., Stability theory of fuzzy differential equations via differential inequalities, *Math. Inequal. Appl.* **2** (1999), no. 4, 551–559.

[63] Lakshmikantham, V. and Liu, X., Impulsive hybrid systems and stability theory, *J. Nonlinear Diff. Eqns.* **5** (1998), 9–17.

[64] Lakshmikantham, V., Matrosov, V., and Sivasundaram, S., *Vector Lyapunov Functions and Stability Analysis of Nonlinear Systems*, Kluwer Academic Publishers, Dordrecht 1991.

[65] Lakshmikantham, V. and McRae, F.A., Basic results for fuzzy impulsive differential equations, *Inequalities and Applications* **4** (2001), 239–246.

[66] Lakshmikantham, V. and Mohapatra, R.N., Basic properties of solutions of fuzzy differential equations, *Nonlinear Studies* **8** (2001), 113–124.

[67] Lakshmikantham, V. and Mohapatra, Ram N., Fuzzy sets and fuzzy differential equations, *Differential Equations and Nonlinear Mechanics* (Orlando, FL, 1999), 183–199, *Kluwer Acad. Publ.*, Dordrecht 2001.

[68] Lakshmikantham, V., Murty, K.N. and Turner, J., Two point boundary value problems associated with non-linear fuzzy differential equations, *Mathematical Inequalities and Applications*, **4** (2001), 527–533.

[69] Lakshmikantham, V. and Trigiante, D., *Theory of Difference Equations: Numerical Methods and Applications*, Academic Press, New York 1988.

[70] Lakshmikantham, V. and Vatsala, A.S., Existence of fixed points of fuzzy mappings via theory of fuzzy differential equations, Fixed point theory with applications in nonlinear analysis, *J. Comput. Appl. Math.* **113** (2000), no. 1-2, 195–200.

[71] LaSalle, J.P., *The Stability of Dynamical Systems*, Regional Conference Series in Applied Mathematics, SIAM, Philadelphia 1976.

[72] Ma, M., Friedman, M., and Kandel, A., Numerical solutions of fuzzy differential equations, *Fuzzy Sets and Systems* **105** (1999), no. 1, 133–138.

[73] Messera, J.L., The meaning of stability, *Bol. Fac. Ingen. Agrimeus Montevideo* **8** (1964), 405–429.

[74] Mizumoto, M. and Tanaka, K., Some properties of fuzzy sets of type 2, *Inform. Control* **31** (1976), 312–340.

[75] Mizumoto, M. and Tanaka, K., Fuzzy sets of type 2 under algebraic product and algebraic sum, *Fuzzy Sets and Systems* **5** (1981), 277–290.

[76] Mohapatra, R. and Zhang, Y., Criteria for boundedness of fuzzy differential equations, *Math. Inequal. Appl.* **3** (2000), no. 3, 399–410.

[77] Nayak, P.C., Oscillation and nonoscillation theorems for second order fuzzy differential equations, *J. Fuzzy Math.* **7** (1999), no. 2, 491–497.

[78] Nayak, P.C. and Nanda, S., Oscillation and nonoscillation theorems of the first and second order fuzzy differential equations, *J. Fuzzy Math* **3** (1995), no. 4, 863–870.

[79] Negoita, C.V. and Ralescu, D.A., *Applications of Fuzzy Sets to System Analysis*, John Wiley, New York 1975.

[80] Neumaier, A., *Interval Methods for Systems of Equations*, Cambridge University Press 1990.

[81] Nguyen, H.T. A note on extension principle for fuzzy sets, *J. Math. Anal. & Appl.* **64** (1978), 369–80.

[82] Nieto, J.., Fuzzy differential equations (Spanish), *XV Congress on Differential Equations and Applications/V Congress on Applied Mathematics, Vol. I, II* (Spanish) (Vigo, 1997), 333-337, Colecc. Congr., 9, Univ. Vigo, Vigo 1998.

[83] Nieto, J.J, The Cauchy problem for continuous fuzzy differential equations, *Fuzzy Sets and Systems* **102** (1999), no. 2, 259–62.

[84] Novak, V., *Fuzzy Sets and Their Applications*, Adam Hilger, Berlin 1989.

[85] Ouyang, H. and Wu, Y., On Fuzzy differential equations, *Fuzzy Sets and Systems* **32** (1989), no. 3, 321–325.

[86] Pal, S.K., *Fuzzy Mathematical Approach to Pattern Recognition*, John Wiley, New York 1987.

[87] Park, D.-G., Kim, M.-H., and Kwun, Y.-C., The existence and uniqueness of fuzzy solution for the nonlinear fuzzy differential equation with nonlocal initial condition, *Far East J. Math. Sci. (FJMS)*, **2** (2000), no. 6, 979–988.

[88] Park, D.-G., Kwun, Y.-C., and Jeong, J.-M., Existence and uniqueness of solutions for fuzzy differential equations, *Far East J. Math. Sci. (FJMS)* **1** (1999), no. 6, 1015–1021.

[89] Park, J.S., Lee, S.Y. and Jeong, J.U., The approximate solutions of fuzzy integral functional equations, *Fuzzy Sets and Systems*, **110** (2000), 79–90.

[90] Park, J.Y. and Han, H.K., Existence and uniqueness theorem for a solution of fuzzy differential equations, *Int. J. Math. Math. Sci.* **22** (1999), no. 2, 271–279.

[91] Park, J.Y. and Han, H.K., Fuzzy differential equations, *Fuzzy Sets and Systems* **110** (2000), no. 1, 69–77.

[92] Park, J.Y., Han, H.K., and Jeong, J.U., Asymptotic behavior of solutions of fuzzy differential equations, *Fuzzy Sets and Systems* **91** (1997), no. 3, 361–364.

[93] Park, J.Y., Han, H.K., and Son, K.H., Fuzzy differential equation with nonlocal condition, *Fuzzy Sets and Systems* **115** (2000), no. 3, 365–369.

[94] Park, J.Y., Jeong, I.H., and Lee, S.Y., Fuzzy differential equations and solution mapping, *J. Fuzzy Math.* **8** (2000), no. 2, 425–432.

[95] Park, J.Y. and Jeong, J.U., A note on fuzzy integral equations, *Fuzzy Sets and Systems*, **108** (1999), 193–200.

[96] Park, J.Y., Lee, S.Y., and Kim, H.M., The existence of solutions for fuzzy differential equations with infinite delays, *Indian J. Pure Appl. Math.* **31** (2000), no. 2, 137–151.

[97] Pearson, D.W., A property of linear fuzzy differential equations, *Appl. Math. Lett.* **10** (1997), no. 3, 99–103.

[98] Puri, M.L. and Ralescu, D.A., Differentials for fuzzy functions, *JMAA* **91** (1983), 552–558.

[99] Puri, M.L. and Ralescu, D.A., Fuzzy random variables, *JMAA***114**(1986), 409–422.

[100] Rådström, H., An embedding theorem for spaces of convex sets, *Proc. Amer. Math. Soc.* **3** (1952), 165–169.

[101] Rizzo, R., Variational Lyapunov method and stability theory of hybrid systems (to appear).

[102] Royden, H.L., *Real Analysis*, Macmillan, London 1968.

[103] Rzezuchowski, T. and Wasowski, J., Differential equations with fuzzy parameters via differential inclusions, *J. Math. Anal. Appl.* **255** (2001), 177–194.

[104] Sakawa, M., *Fuzzy Sets and Interactive Multiobjective Optimization*, Plenum Press 1993.

[105] Seikkala, S., On the fuzzy initial value problem, *Fuzzy Sets and Systems*, **24** (1987), 319–330.

[106] Siljak, D.D., *Large Scale Dynamic Systems*, North-Holland, New York 1978.

[107] Song, S. and Wu, C., Remark on approximate solutions, existence, and uniqueness of the Cauchy problem of fuzzy differential equations, *J. Fuzzy math.* **6** (1998), no. 4, 923–928.

[108] Song, S. and Wu, C., Existence and uniqueness of solutions to the Cauchy problem of fuzzy differential equations, *Fuzzy Sets and Systems* **110** (2000), no. 1, 55–67.

[109] Song, S., Guo, L., and Feng, C., Global existence of solutions to fuzzy differential equations, *Fuzzy Sets and Systems* **115** (2000), no. 3, 371–376.

[110] Song, S., Liu, Q.-Y. and Xu, Q.-C., Existence and comparison theorems for Volterra fuzzy integral equations in (E^n, D), *Fuzzy Sets and Systems*, **104** (1999), 315–321.

[111] Taylor, A.E., *Introduction to Functional Analysis*, Wiley, New York 1964.

[112] Vorobiev, D. and Seikkala, S., Towards the theory of fuzzy differential equations, Preprint, University of Oulu (1999), 16 pp.

[113] Weaver, W., Science and complexity, *American Scientist* **36** (4) (1948), 536–544.

[114] Wu, C. and Song, S., Existence theorem for the Cauchy problem of fuzzy differential equations under compactness-type conditions, *Inform. Sci.* **108** (1998), no. 1-4, 123–134.

[115] Wu, C., Song, S., and Lee, E.S., Approximate solutions, existence, and uniqueness of the Cauchy problem of fuzzy differential equations, *J. Math. Anal. Appl.* **202** (1996), no. 2, 629–644.

[116] Zadeh, L.A., Fuzzy sets, *Inf. Control* **8** (1965), 338–353.

[117] Zadeh, L.A., Similarity relations and fuzzy orderings, *Inf. Sci.* **3** (1971), 177–206.

[118] Zadeh, L.A., The concept of linguistic variable and its applications to approximate reasoning, *Inf. Sci.* **9** (1975), 43–80.

[119] Zhang, Y. and Wang, G., Time domain methods for the solutions of n-order fuzzy differential equations, *Fuzzy Sets and Systems* **94** (1998), no. 1, 77–92.

[120] Zhang, Y., Qiao, Z., and Wang, G., Solving processes for a system of first- order fuzzy differential equations, *Fuzzy Sets and Systems* **95** (1998), no. 3, 333–347.

[121] Zhang, Y., Wang, G., and Liu, S., Frequency domain methods for the solutions of n-order fuzzy differential equations, *Fuzzy Sets and Systems* **94** (1998), no. 1, 45–59.

[122] Zhou, Z. and Yu, Z.X., Minimal and maximal solutions of fuzzy differential equations (Chinese), *Mohu Xitong yu Shuxue* **9** (1995), no. 4, 54–59.

[123] Zimmerman, H.J., *Fuzzy Set Theory and its Applications*, Kluwer-Nijhoff Publishing, 1984.

Index

Milton Keynes UK
Ingram Content Group UK Ltd.
UKHW052017071024
449327UK00027B/2307